Taxonomy Matching Using Background Knowledge

Heiko Angermann · Naeem Ramzan

Taxonomy Matching Using Background Knowledge

Linked Data, Semantic Web
and Heterogeneous Repositories

 Springer

Heiko Angermann
University of the West of Scotland
Paisley
UK

Naeem Ramzan
University of the West of Scotland
Paisley
UK

ISBN 978-3-319-89157-6   ISBN 978-3-319-72209-2   (eBook)
https://doi.org/10.1007/978-3-319-72209-2

Printed on acid-free paper

This Springer imprint is published by Springer Nature
The registered company is Springer International Publishing AG
The registered company address is: Gewerbestrasse 11, 6330 Cham, Switzerland

Preface

Taxonomies and its related technologies ontologies, schemas, directories, and electronic catalogs are using a hierarchical order to model a field of interest in a formal way and have become ubiquitous across domains and information management systems during the last years. Those are crucial to categorize customers according to their accompanying branch inside a Customer Relationship Management (CRM) system, are used in Product Information Management (PIM) systems to categorize goods of an enterprise according to the most related category, are necessary to classify the to the goods belonging assets inside a Media Asset Management (MAM) system, or are used in e-commerce systems to help customers finding the desired products.

At the same time, enterprises have the need to exchange data between the systems inside a complex data warehouse scenario, but also to exchange data between other providers. For example, the interaction between CRM and PIM can be used to filter for relevant customers, when setting up a new marketing strategy for specific branches. Or, when the own products and services have to be distributed in global marketplaces like Amazon or eBay, instead of only using the own marketplace, to increase sale. However, as the single systems and different providers usually use their own taxonomy to classify the instances according to the interpretation and embossing of the domain, the exchange is significantly made much more difficult. Correspondences according to the semantic similarity between the categories, i.e., more precisely concepts, have to be detected, to allow a (semi)-automatic exchange between the databases.

Aim of the Book

During last years, an increasing number of algorithms and systems has been proposed, which aim to find correspondences between taxonomies. Such taxonomy matching algorithms and systems detect the similarity between the taxonomies laborious or facile depending on the differences between the taxonomies (i.e., taxonomic heterogeneity), and depending on the utilized matching techniques. In contrast to previous years, recent approaches now use an interaction between different matching techniques inside one matching strategy to overcome different types of heterogeneity inside one framework. In addition, recent attempts make use of various resources of background knowledge, to further help inferring similarities between the repositories, instead of merely using the knowledge derived from the two taxonomies. Both, the improved matching strategies and the usage of flexible background knowledge, have proven a significant improvement in the field according to various dynamic, as well as static benchmarks.

The book at hand is anticipated to be an assortment of existing and new matching techniques, a review of new introduced matching algorithms, an analysis of recently provided matching systems, as well as a comparison between matching evaluation approaches. In addition, it will investigate the different attempts according to the most important evaluation campaign in the field, namely the Ontology Alignment Evaluation Initiative (OAEI). Through this, the principal aspiration of this book is not only to cover the theoretical and practical fundamentals, but also to cover the state-of-the-art techniques and challenges of the subject area. A number of promising developments and innovative guidelines will also be explained in this book with the aim of motivating and informative opportunity for academic research and practice. In short, the following four objectives are followed by the book at hand:

- The main objective of this book is to provide an in-depth knowledge about the state of the art in the field of taxonomy matching, including its related fields ontology matching, and schema matching.
- It will address new generation matching attempts, including matching strategies, matching algorithms, matching systems, OAEI campaigns, as well as alternative evaluations.
- It will confer the way in which the recent and future attempts make use of different sources of background knowledge to allow a more precise matching between the repositories.
- It will cover research aspects, but also aspects to practitioners that make use of the matching approaches.

Outline of the Book

Part I is dedicated to the definition and motivation of the problem taxonomic heterogeneity. For this part, two chapters are included. The background to the research field is given in Chap. 1. Hereby, the problem of taxonomy matching is explained in detail. In addition, a categorization of works is provided. Afterwards, the different types of taxonomic heterogeneity are detailed. Finally, the main techniques for evaluating recent works are explained using also Chap. 1. A more fine-granular differentiation between the four different types of heterogeneity is provided in Chap. 2. Hereby, a novel methodology is presented to clarify when and how taxonomies can be matched laborious or facile. Afterwards, the methodology is applied to terminological heterogeneity, conceptual heterogeneity, syntactical heterogeneity, as well as semiotic heterogeneity.

Part II is focusing on the works aiming to overcome taxonomic heterogeneity, and the progress made in the field. For this part, two chapters are included. Chapter 3 is reviewing the different types of matching techniques existing, before the book is reviewing the most recent matching algorithms. Hereby, algorithms focusing on one of the four types of heterogeneity are included. It is reviewed, in which techniques are used and combined, as well as, which sources of background knowledge are used. Finally, matching systems are reviewed that are able to overcome different types of taxonomic heterogeneity using a more comprehensive matching strategy. The systems included in Chap. 3 have all participated in different evaluation campaigns and have proven to outperform existing systems. Using Chap. 4, the most widely used evaluation methodology is reviewed. Hereby, the campaigns of the last 5 years are discussed in detail.

Part III is discussing related research fields making use of taxonomic heterogeneity. For this part, one chapter is included. In Chap. 5, related research fields aiming to evolve and adapt taxonomies according to insights and interaction are reviewed. This includes the fields of dynamic taxonomies, personalized directories, and catalog segmentation. For each field, the background is explained, concrete applications are discussed, as well as recent approaches are reviewed. Finally, the main techniques for evaluating works of those fields are explained. Afterwards, the research field making use of taxonomies to analyze preferences is reviewed, i.e., the field of recommender systems. Hereby, the different types of recommender systems making use of taxonomies are discussed.

Part IV concludes the book. Chapter 6 is summarizing the findings of the book and discussing the recent state of taxonomic taxonomy matching.

Readership and Lecture Guide

The target audience of this book will be composed of professionals and researchers working in the fields of taxonomy matching and its related fields ontology and schema matching in various disciplines, especially involved with artificial and business intelligence. The book should also serve as a solid and advanced-level course supplement to taxonomy matching for upper-level undergraduate and postgraduate students studying this subject. In addition, the book will offer insights and support to groundbreaking and innovative studies apprehensive with the development of taxonomy matching approaches within different types of research and working communities. In short, the target audience is as follows:

- Researchers with a focus on the fields of taxonomy matching, including taxonomy mapping and merging, or with a focus on the field of semantic similarity assessment.
- Lectures giving courses in the disciplines of information management, including the subdisciplines of electronic commerce, master data management, as well as content management.
- Graduated who specialize in the area of metadata management, including the techniques of formal metadata management as well as informal metadata management.
- Practitioners who specialize in the areas of data processing and analyzing, including the subareas of data science and business intelligence.

Paisley, UK   Heiko Angermann
August 2017   Naeem Ramzan

Contents

Acronyms

AML	Agreement Maker Light
API	Application Programming Interface
B2B	Business to Business
B2C	Business to Consumer
CPU	Central Processing Unit
CRM	Customer Relationship Management
CSV	Comma Separated Values
DAML	DARPA Agent Markup Language
DOID	Human Disease Ontology
FCA	Formal Concept Analysis
GOMMA	Generic Ontology Matching and Mapping Management
ICTCLAS	Institute of Computing Technology Chinese Lexical Analysis System
ID3	Iterative Dichotomiser 3
JDBC	Java Database Connectivity
KRSS	Key Registration Service Specification
LIPS	Logical Inferences Per Second
LogMap	Logic-based and Scalable Ontology Matching/Mapping
MeSH	Medical Subject Headings
MWE	Multi Word Expression
NAICS	North American Industry Classification System
NAPCS	North American Product Classification System
OAEI	Ontology Alignment Evaluation Initiative
OBO	Open Biological and -medical Ontologies
OM-Reasoner	Ontology Matching Reasoner
OWL	Web Ontology Langauge
PIM	Product Information Management
RDF	Resource Description Framework
RiMOM-IM	Risk Minimization based Ontology Mapping—Instance Matching
SAX	Simple API for XML

SKOS	Simple Knowledge Organization System
SPARQL	SPARQL Protocol And RDF Query Language
SQL	Structured Query Language
TF–IDF	Term Frequency–Inverse Document Frequency
Turtle	Terse RDF Triple Language
Uberon	Uber Anatomy Language
XMap++	Extensible Mapping ++
XML	Extensible Markup Language
YAM++(-)	(Not) Yet Another Matcher ++(-)

Part I
Introduction to Taxonomy Matching

Chapter 1
Background Taxonomy Matching

Abstract During the last decades, the amount of data has increased dramatically. This is because enterprises are using various information management systems, and because of the nowadays goal for interlinking between such systems to gain new information. To effectively store the extensive amount of data in a structured way, two metadata paradigms are used predominantly: taxonomies (formal metadata) and folksonomies (informal metadata). Taxonomies are classifying objects based on hierarchically ordered formal concepts. Because of this, taxonomies have its benefits for controlling how instances can be classified. However, when exchanging data across multiple information systems inside a single firm, or with external systems (e.g., digital marketplaces), the underlying taxonomies are very often not the same. This is because the domain is different or because the underlying methodologies are varying. Logically, the underlying taxonomies have to be mapped before exchanging data in a proper way, named taxonomy matching. Providing the chapter at hand, a detailed overview of this research area is given, including an explanation of its principles, the aim of matching taxonomies, the problem of heterogeneity, a categorization for matching attempts, as well as an overview of the mainly used evaluation metrics.

During the last decade, the amount of data to be stored over different databases, and the amount of information to be handled over various information management systems, has increased dramatically [9]. In e-commerce for example, the online marketplaces provide an extensive number of various products and services to their customers, and the commercialization is done using different online and offline marketing strategies. In addition, the customers can also use various channels and devices to enter the multichannel marketplaces that interact with other marketplaces or systems [27].

Two metadata paradigms, i.e., data about data, have arisen during the last centuries to structure data inside information management systems. On the one hand, the keyword-based method called folksonomy describes information in the form of informal tags [142]. Those tags are lightweight, human understandable and offer the possibility to create interlinked networks. However, as there are no restrictions for tagging information, the tags contain semantic ambiguities and synonyms [101]. On the other hand, to model a field of interest in a formal way, the second method called

© Springer International Publishing AG 2017

H. Angermann and N. Ramzan, *Taxonomy Matching Using Background Knowledge*, https://doi.org/10.1007/978-3-319-72209-2_1

taxonomy is used. Taxonomies, also called directories, and in e-commerce named
e-catalogs, are subcategories of ontologies, which are using hierarchically ordered
concepts to model a field of interest in a formal way [67]. This hierarchical representation of a domain has its merits for navigation and for exploring similar items [142].
For example, to categorize customers according to their accompanying branch inside
a Customer Relationship Management (CRM) system, to categorize goods according
to categories inside a Product Information Management (PIM) system, to classify
assets inside a Media Asset Management (MAM) system, in E-Commerce systems
to help customers finding the desired products, or in Enterprise Resource Planning
(ERP) systems to structure product master and lifecycle data.

However, as nowadays there is often a need to combine, exchange, and interact
data over different systems and channels, there is often a need to compare the two
data repositories based on the underlying taxonomy, for example, if a retailer has
an own online retailing platform but also wants to distribute the products or services on a global marketplace as provided by Amazon[1] or eBay.[2] However, as most
of the enterprises are using their own taxonomy to model over hundreds of interrelated concepts, a manual comparison between two data repositories would be a
time-intensive and error-prone task. To (semi)-automatically detect matches between
two taxonomies, a broad research community is treating the paradigm of *Taxonomy
Matching* and *Ontology Matching*. Approaches introduced in this research field find
correspondences between formal structured concepts laborious or facile, depending
on the similarity and dissimilarity existing between the taxonomies, named *Taxonomic Heterogeneity*. According to the literature, four types of heterogeneity exist,
whereby one or multiple types of heterogeneity can exist between two taxonomies
[157]: terminological heterogeneity (different labels/languages), conceptual heterogeneity (contradictory structures), syntactical heterogeneity (varying data models),
and semiotic heterogeneity (disparate cognitive interpretations).

Because the type(s) of heterogeneity existing between two taxonomies decisively
affect finding correspondences, recent matching approaches are differing from the
approaches published in the century before, in two directions. Firstly, recent attempts
are focusing on the combination of multiple techniques, instead of using a single
technique. This allows that different types of heterogeneity can be overcome using
a single approach, and the matching quality result is usually increased. Secondly,
recent attempts are using so-called background knowledge. Background knowledge
in the form of lexicons, thesauri, or additional taxonomies being published as linked
data or elsewhere is additional resources used to help inferring further relationships
between concepts and thus helps assessing similarity between concepts/taxonomies
[157]. Through this, the matching quality result is highly increased, as the amount of
information to be used for analyzing the concepts grows by every resource of background knowledge used. The latest evaluations performed in the field evidenced that
the taxonomy matching systems perform better than more resources of background
knowledge they are using [35]. To understand the core principles of taxonomies, the

---

[1]http://www.amazon.com.
[2]http://www.ebay.com.

aim of taxonomy matching, and the problem of taxonomy heterogeneity, this chapter is used describing those problems in detail.

The remainder is organized as follows. In Sect. 1.1, the principles of taxonomies are explained. This includes a definition of the term as well as the different types of concepts included. In Sect. 1.2, the research field of taxonomy matching is discussed. It details the aim of the works introduced in this field, and it describes the main steps being required to match two taxonomies. In Sect. 1.3, the problem of taxonomic heterogeneity is introduced. Hereby, an explanation of the four types of heterogeneity is given. Based on the before-gone sections, a categorization of works is presented using Sect. 1.4. The main methodologies and metrics to evaluate matching approaches are introduced in Sect. 1.5. Finally, this chapter concludes in Sect. 1.6.

1.1 Taxonomy Principles

A *Taxonomy* (Θ), also named directory, schema, and in e-commerce referred to as e-catalog, is subcategories of ontologies. Those are describing a domain of objects with similar properties inside an out-tree, as given in Fig. 1.1. Contrary to ontologies, a taxonomy is only describing hierarchical relationships (hypernym, hyponym), but not arbitrary complex relationships (meronyms, antonyms, synonyms) [130, 142], with (see Eq. 1.1):

$$\Theta = (\{\Phi\}, \{\Lambda\}), \tag{1.1}$$

which is using a set of concepts Φ for describing terms with a label, i.e., name of the concept, and a set of edges Λ connecting less general with more general concepts of different levels. The edges between the concepts represent the hierarchical relationships inside the taxonomy. For example, a taxonomy consisting of three hierarchically ordered levels utilizes a root concept as the most general concept, different super concepts detailing a root concept, and sub concepts detailing the super concept, which is in turn, a sub concept of the root concept (see Fig. 1.1a and b).

A single concept ϕ_C is a *Sub Concept*, formally $subof$, if it is a less generalized concept of another concept, ϕ_B, as given in Eq. (1.2), if:

$$\phi_C = subof(\phi_B) :\Leftrightarrow (\phi_C \subset \phi_B) \wedge ((\phi_C \wedge \phi_B) \in \Phi), \tag{1.2}$$

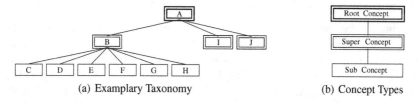

(a) Examplary Taxonomy   (b) Concept Types

Fig. 1.1 Hierarchical structure of an exemplary taxonomy and its concept types

where ϕ_C and ϕ_B are two concepts of taxonomy Θ described through Φ and Λ. This relationship is also referred to as *is-a* relationship. Consequently, a *Super Concept* ϕ_B, formally $superof$, is a more generalized concept of ϕ_C, as given in Eq. (1.3), if:

$$\phi_B = superof(\phi_C) :\Leftrightarrow \phi_C = subof(\phi_B). \qquad (1.3)$$

A *Sibling Concept* ϕ_D of ϕ_C, formally $sibof$, is the relationship between two concepts sharing the same super concept, as given in Eq. (1.4), if:

$$\phi_D = sibof(\phi_C) :\Leftrightarrow (\phi_D \wedge \phi_C) = subof(\phi_B). \qquad (1.4)$$

A *Root Concept* ϕ_A, formally $rootof$, is a concept that has no super concept, as given in Eq. (1.5), in which:

$$A = rootof(\Theta) :\Leftrightarrow \nexists superof(\phi_A). \qquad (1.5)$$

Besides the label, each concept can have an optional description (e.g., "A …used for …"), and a set of optional properties acting as additional metadata (e.g., "Color"). The creation of the taxonomy is either performed through expert(s) knowing the technical details of the entities belonging to a concept, or by matching to formal resources, e.g., to a standard taxonomy, which provides predefined sets of concepts for specific domains. The **W**eb **O**ntology **L**angauge (OWL) and the **R**esource **D**escription **F**ramework (RDF) are the proprietary used semantic data languages to store such taxonomic relationships (for further details see [26, 109]). SPARQL **P**rotocol **a**nd **R**DF **Q**uery **L**anguage (SPARQL) is the mainly used language to query against the taxonomies [139]. Such languages are all based on **E**xtensible **M**arkup **L**anguage (XML), a programming language for managing data stored inside a hierarchical database system describing entities with the help of markups. To construct taxonomies, the authors in [87] defined three tasks:

1. *Building the Taxonomy.* Either through a bottom-up approach, i.e., combination of sub concepts, or with a top-down approach, i.e., splitting of super concepts.
2. *Grouping the Concepts.* The grouping of sibling concepts is predominantly achieved by referring to background knowledge resource(s). For example to a standard taxonomy, or the semantic lexicon WordNet, which is the most widely used resource [58].
3. *Assigning the Super Concepts.* The last task aims to determine super concepts for the grouped sub concepts, i.e., the is-a relationships.

During the last decade, taxonomies have arisen to be an essential part in various information management applications. For example, taxonomies are used on e-commerce sites and are implemented in all recent e-commerce applications as e-catalogs, as well as in the included **P**roduct **I**nformation **M**anagement (PIM) components for managing products, and in the included **C**ustomer **R**elationship **M**anagement (CRM) components for managing customers [8, 133]. The e-catalogs, along with the concepts, in e-commerce named categories, are utilized for supporting

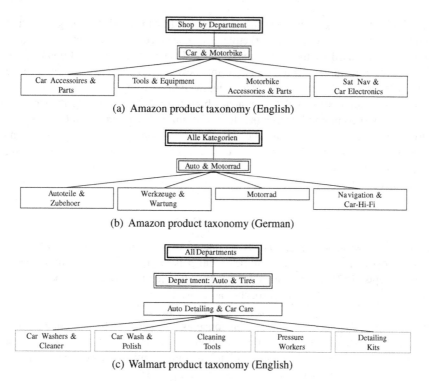

(a) Amazon product taxonomy (English)

(b) Amazon product taxonomy (German)

(c) Walmart product taxonomy (English)

Fig. 1.2 Three exemplary subsets of real-world e-catalogs in e-commerce

the user when navigating through the marketplace, to make selections in it, and to place orders; see Fig. 1.2. The labels of e-catalogs mainly consist of a combination of different word sequences, i.e., **M**ulti **W**ord **E**xpression (MWE), which differ according to the used language (see Fig. 1.2a and b) and according to the marketplace (see Fig. 1.2a and c). MWEs are defined as idiosyncratic interpretations that cross word boundaries (or spaces) [150]. For example, the Amazon concept "Car & Motorbike" includes four less generalized concepts that all use a label consisting of MWEs. Through this, each concept combines semantically less or more similar sequences as a label.

1.2 Matching Taxonomies

Nowadays there is often a need to combine, exchange, and interact data over different systems, channels, and providers, and data has to be compared. For example, if a retailer has an own online retailing platform, but also wants to distribute the products or services on global marketplace. Such an exchange between products that should be published on different marketplaces is usually achieved through comparing the underlying taxonomies. Through this, it is ensured that the products being classified into own category of the source marketplace are classified into a semantically similar category being used by the target marketplace. However, as a manual comparison between two repositories would be a time-intensive and error-prone task, works have been introduced supporting and/or handling the comparison.

To (semi)-automatically detect matches between two heterogeneity resources, a broad research community is treating the paradigm of **Taxonomy Matching**. Works introduced in this paradigm have the aim of finding correspondences between concepts of two taxonomies by using a so-called matching strategy during a matching operation, also referred to as matching process [132]. The resulting alignment of the matching operation is captured in a four-tuple; see Eq. 1.6:

$$\Theta_A, \Theta_B = (\alpha, \phi_{A_A}, \phi_{A_B}, \rho), \tag{1.6}$$

where α is a local identifier, ϕ_{A_A} is a concept of the source taxonomy Θ_A, ϕ_{A_B} is a concept of the target taxonomy Θ_B, and ρ is the theoretical relation between both concepts [157]. The relationship uses subsumption ($\subsetneq, \supsetneq$), equivalence ($\equiv$), or disjointness ($\sqcup$) [14]. Recent approaches add a truth-value in the form of a weighted Boolean, i.e., a data type stating if a test is true or false. The truth-value states how sure the theoretical relation between the concepts is, whereby a higher value usually represents a higher likelihood that the relation is true [157]. Such systems, without considering the output, input, and verification components, produce the alignments (matching operation) in a four-step process (see Fig. 1.3):

1. *Selecting Matching Strategy*. A matching strategy can include various element-level and/or structure-level matching techniques. The strategy should be chosen depending on the disparity between the two taxonomies.
2. *Performing Matching Strategy*. The matching process to detect correspondences makes use of background knowledge in the form of lexicons, formal, or informal sources, to help inferring the semantics and relationships between taxonomies [55].

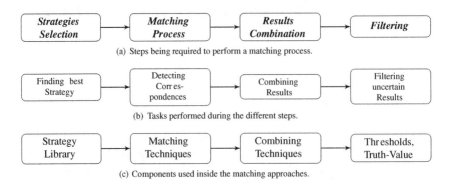

(a) Steps being required to perform a matching process.

(b) Tasks performed during the different steps.

(c) Components used inside the matching approaches.

Fig. 1.3 Matching Operation illustrated as a four-step process

3. *Combining Matching Results*. The results obtained through different matching techniques must be combined to state the final similarity. Recent approaches add a similarity matrix at this step to rank the different results obtained for identical relationships.
4. *Filtering Matching Results*. Filtering irrelevant matchings with a threshold applied to the correspondence and its truth-value. The former value is compared with the similarity value to state if a similarity exists, or not. The latter value is added to the correspondences, to state how sure the theoretical relation between the concepts is.

1.3 Taxonomic Heterogeneity

The quality result performed through a matching operation is significantly affected by the cognitive and methodical disparity between two taxonomies, named ***Taxonomic Heterogeneity***. For example, when the marketing expert of Walmart, as given in Fig. 1.2c as source taxonomy Θ_A, wants to order price lists from a German supplier producing sparkling plugs, as given in Fig. 1.2b as target taxonomy Θ_B. The differences between both taxonomies concern the languages used to define the labels of the concepts, and the number of concepts used to model the domain. The authors in [55] outlined four categories of heterogeneity to detail the differences between taxonomies:

- ***Terminological Heterogeneity*** appears when the labels of concepts are different [54]. Either because of different languages are in use, e.g., Θ_A in English, and Θ_B in German, when using different technical sublanguages, or when using synonyms.
- ***Conceptual Heterogeneity*** arises if two taxonomies are using different models [55]. The taxonomies are representing the domain with different axioms, i.e., true statements, or different concepts [54]. For example, the concept "Auto Detailing & Car Care" has five sub concepts, whereas the concept "Auto & Motorrad" has only four sub concepts, but some of the concepts are semantically similar.
- ***Syntactical Heterogeneity*** occurs when different data languages/models are used to store the taxonomies [54]. For example, Θ_A is stored in OWL, but Θ_B is stored in RDF. As OWL has a precise mapping to RDF, thus is comparable with arbitrary RDF graphs. If languages are using a different knowledge representation formalism, matching approaches are required to translate between the schemata.
- ***Semiotic Heterogeneity*** emerges when persons misinterpret concepts, respectively, the is-a relationships between the concepts. For example, a customer does not expect "Pressure Workers" to detail "Auto Detailing & Car Care", as this tool can also be used for cleaning houses and gardens.

1.4   Matching Categorization

Since the advent of research field taxonomy matching, a lot of work has been introduced. However, the works differ mainly in four directions. Firstly, the methodology
of how similarity between taxonomies, respectively correspondences between the
single concepts, is detected. Secondly, which type of heterogeneity should be overcome. Thirdly, if the work is treating a single type of heterogeneity, or if multiple
types of heterogeneity are treated. And fourthly, if the work is focusing on improving
the quality matching results, or if is aimed to evaluate matching attempts.

   To better distinguish between matching approaches, recent literature applies four
categories, based on the before-gone differentiations [55, 126, 157].

- *Matching Techniques* are single element-level and structure-level techniques used
  in matching algorithms and systems.
- *Matching Algorithms* aim to overcome heterogeneity by utilizing one concrete matching strategy including multiple techniques. In this review, algorithm
  approaches are included, which provide a matching strategy that has not been presented previously and that refers to different directions regarding progress evident
  by the Ontology Alignment Evaluation Initiative (OAEI) campaigns.
- *Matching Systems* instead aim to overcome multiple types of heterogeneity by
  providing flexible matching strategies. Hereby, systems are included that have
  successfully participated in recent OAEI campaigns. The systems differ in the
  kind of processing the correspondences, concrete matching techniques used, and
  the sources of background knowledge utilized.
- *Matching Evaluations* help assessing the available matching systems and algorithms [126]. This category includes approaches describing metrics, campaigns,
  benchmarks, and datasets to evaluate the matching systems. This category includes
  an in-depth analysis of the latest OAEI campaigns, but also alternative evaluation
  approaches including benchmarks, datasets, and metrics.

1.5   Matching Evaluation

In the field of taxonomy matching, *Matching Evaluation* approaches are utilized
to assess the implemented matching strategy against comparable challenges. The
works on taxonomy matching are following a uniform evaluation scenario. Along
with standard datasets provided for evaluating matchers, the standard metrics in
information retrieval (i.e., re-establishment of information) are applied, which are
the following: Precision, Recall, and F-Measure [157]. For evaluating semantic similarity between two concepts of one single taxonomy, e.g., for evaluation sources
of background knowledge, the accuracy of the techniques is analyzed. However, if
the dataset is imbalanced, the comparison of Sensitivity and Specificity is taken into
account, resulting the Balanced Accuracy.

Formally, **Precision** is the measure of correctness (see Eq. 1.7):

$$Precision = \frac{\sum True\ Positive}{\sum True\ Positive + \sum False\ Positive}, \tag{1.7}$$

where $True\ Positive$ is a relevant class identified as relevant, and $False\ Positive$ is irrelevant class identified as relevant. Contrary, the **Recall** is a measure of completeness (see Eq. 1.8)

$$Recall = \frac{\sum True\ Positive}{\sum False\ Negative + \sum True\ Positive}, \tag{1.8}$$

where $False\ Negative$ is a relevant class identified as non-relevant. The harmonic mean of both before-mentioned scores is named **F-Measure** (see Eq. 1.9)

$$F - Measure = 2 \times \frac{Precision \times Recall}{Precision + Recall}. \tag{1.9}$$

The **Accuracy** measures systematical errors, also named hit rate (see Eq. 1.10)

$$Accuracy = \frac{\sum True\ Positive + \sum True\ Negative}{\sum True\ Positive + \sum False\ Positive + \sum False\ Negative + \sum True\ Negative} \tag{1.10}$$

where can be seen that now the $True\ Negative$ statements are also taken into account, i.e., a irrelevant class identified as non-relevant. As for imbalanced datasets, this measure is not meaningful, and the **Balanced Accuracy** is used, with (see Eq. 1.11):

$$Balanced\ Accuracy = \frac{Sensivity + Specificity}{2}, \tag{1.11}$$

where $Sensitivity$ is the proportion of correctly identified positives, and $Specificity$ is the proportion of correctly identified negatives. The **Sensitivity** is the rate of true positive statements (see Eq. 1.12), and **Specificity** is the rate of true negative statements (see Eq. 1.13):

$$Sensitivity = \frac{\sum True\ Positive}{\sum True\ Positive + \sum False\ Negative}. \tag{1.12}$$

$$Specificity = \frac{\sum True\ Negative}{\sum True\ Negative + \sum False\ Positive}. \tag{1.13}$$

As datasets, the datasets provided by OAEI are usually taken, which are not taxonomies, but more expressive ontologies. OAEI is in the form of providing ontology/taxonomy-specific tracks and test cases to evaluate taxonomy matching systems on the same basis for drawing conclusions regarding the best matching strategy. As it is held every year to provide a comparable evaluation regarding the improvement of existing and new matching systems, it is the most important matching evaluation in the field of taxonomy matching. Through comparing the included matching strategies it can be stated, which matching techniques should be utilized and combined to increase matching quality accuracy, which sources of background knowledge can be used to help assessing semantic similarity, if user involvement (i.e., involving the (non-)expert during the computation process) can increase accuracy, how the user involvement should be provided, etc. The specific tracks are as follows:

- The *Ontology Track* provides the conference test case, which consists of 16 taxonomies describing the domain of organizing conferences. This track can be used to evaluate taxonomy matching strategies according to the overcoming of terminological and conceptual heterogeneity. The best systems according to its F-Measure scores are the following: YAM++, followed by AML.
- The focus of the *Multifarm Track* is to confront the systems with the case of multilingualism. The problem is divided into two subproblems: matching taxonomies of the same domain of interest and taxonomies of different domains. This track uses the *OntoFarm* dataset. It includes seven taxonomies describing concepts of conferences. These taxonomies were translated in eight different languages (French, German, Dutch, Portuguese, Spanish, Czech, Chinese, and Russian), based on the English version. The two best systems for the first task are the following: AML and LogMap. For the second task, the best system is MapSSS [33], followed by AML.
- The *Directories and Thesauri Track*, more precisely, its library test case, aims at providing a matching task between the *Thesaurus Sozial-wissenschaften* (TheSoz)[3] and the *Standard Thesaurus Wirtschaft* (STW)[4] thesaurus. STW represents an economical science thesaurus, whereas TheSoz depicts a social science thesaurus. A remarkable progress has been made during the last five years according to its F-Measure score: +0.10 was resulted by AML (0.80).
- The goal of the *Interactive Matching Track* is to show if user interaction can improve the matching result. For this track, the *biblio* dataset is used. The two best systems of the conference test case resulted a remarkable improvement when involving the expert user during the matching task (YAM++: +16,19%, and AML: +14.05%). No progress has been made during 2013 and 2015. Until now, this track still only considers the expert, not the non-expert user. However, OAEI has started to treat this challenge with included error rates.
- Other test cases are the *Large Biomedical Ontologies* with its focus on very large biological taxonomies, *Ontology Alignment for Question Answering* used to evaluate the exploitation of correspondences, and the *Anatomy* test case with its focus on biomedical ontologies.

---

[3]http://lod.gesis.org.
[4]http://zbw.eu/stw.

Besides OAEI, other attempts exist to evaluate taxonomy matching approaches. On the one side, there are benchmarks centering on a comparable evaluation of the systems. As measuring the quality of various matching approaches is still a difficult task, these theories aim to create uniform metrics for measuring the matching quality and matching efficiency. On the other side, datasets are proposed to compare the different systems to identical problems because of the matching quality strongly depends on the input taxonomies.

A benchmark approach named *MostoBM* was introduced by [147]. Their approach provides a catalog of real-world data exchange patterns. Their patterns pay attention to terminological, syntactical, and conceptual heterogeneity. Furthermore, the authors determined two kinds of parameters for evaluating matchers: structure, which means the number of levels, and data parameters, which means related concepts and data properties that are standing for the number of individuals, data properties, and object properties.

Reference [37] proved that the standard metrics (*Error Rate*, *Balanced Error Rate*, *Average Hierarchical Loss*, and *F-Measure*, for details see [138]) do not value the effects of incorrect matchings. Their proposed algorithm takes the matching accuracy and the assessment of revenue loss, which is resulted through incorrectly matchings, into account. It weights the sum of classification errors and the significance of the assigned instances. The approach is limited to the retailer domain.

For evaluating a wide range of mapping tasks in relation to ontology data integration, the authors in [137] presented the RODI benchmarking suite. Recently, the suite consists of three conference ontologies provided through OAEI, and the created corresponding relational databases. Three different types of challenges are considered inside the suite: naming conflicts, structural heterogeneity, and semantic heterogeneity. The former considers the use of different conventions to name the concepts either through using identifiers or abbreviations instead of technical artifacts, or through using synonyms (i.e., terminolgical heterogeneity). The structural/conceptual heterogeneity challenge includes differences because of the different modeling constructs (type conflicts), differences because of the representation inside the schema (key conflicts), and homogeneity because of different dependencies (i.e., 1:1, 1:N, N:N; dependency conflicts). The latter considers the differences because of an impedance mismatch and semantic expressiveness.

1.6 Conclusions

This chapter has provided the background to the research field taxonomy matching.

The chapter started with explaining the principles and different concept types used in taxonomies. Afterward, the aim of taxonomy matching was discussed, before the problem of taxonomic heterogeneity was discussed. Afterward, a categorization framework for works on taxonomy matching was presented. Finally, the metrics and datasets for evaluation matching frameworks have been explained.

Chapter 2
Background of Taxonomic Heterogeneity

Abstract As nowadays, often different systems are used and interlinked, very often the underlying taxonomies are heterogeneous. Taxonomic heterogeneity is the cognitive and methodical disparity between two taxonomies describing the same domain of interest and occurs in four categories: terminological heterogeneity (different languages), conceptual heterogeneity (contradictory structures), syntactical heterogeneity (varying data models), and semiotic heterogeneity (disparate cognitive interpretations). Taxonomy matching systems and algorithms find correspondences between concepts laborious or facile based on the type(s) of heterogeneity between taxonomies. Because of this, it is not sufficient to reduce the observation of inhomogeneity on merely the categories of heterogeneity only, as for every type of heterogeneity different grades of unevenness exist. Based on indicated various degrees of heterogeneity a new methodology is established using this chapter. The degree of heterogeneity depends on the used semantic, syntax, and fine granularity, to express an identical domain of interest. After reviewing the existing classification, the novel methodology is discussed, before it is directly applied on recent taxonomy matching systems and compared with results provided through OAEI. The experimental evaluation highlights the efficiency of the proposed methodology.

Taxonomic heterogeneity is the cognitive and methodical disparity between two taxonomies describing the same domain of interest. The heterogeneity itself is categorized into four subjects according to existing literature [157]: terminological heterogeneity (different languages, and language usage), conceptual heterogeneity (contradictory structures, and levels), syntactical heterogeneity (varying data models), and semiotic heterogeneity (disparate cognitive interpretations). As most of the enterprises are using their own taxonomy, even if those are describing an identical domain, a manual comparison between two heterogeneous taxonomies would be a time-intensive task. To (semi)-automatically detect correspondences between taxonomies, a remarkable research community is addressing the paradigm of *Taxonomy Matching*.

Matching approaches aim to find correspondences between two heterogeneous taxonomies by using a matching strategy. Of course, the approaches find those correspondences laborious or facile, depending on the heterogeneity existing between

© Springer International Publishing AG 2017

H. Angermann and N. Ramzan, *Taxonomy Matching Using Background Knowledge*, https://doi.org/10.1007/978-3-319-72209-2_2

the two formal structure repositories. As the heterogeneity between two taxonomies decisively affects finding correspondences by a matching algorithm or system, the differentiation into merely four types of heterogeneity can be criticized. For example, in the case of terminological heterogeneity, some frameworks are limited to match taxonomies expressed in same language, whereas others are utilizing a translating machine to be able for matching taxonomies expressed in different languages. Thus, the latter-mentioned approaches can overcome terminological heterogeneous taxonomies differing in the used vocabulary to express the taxonomies. As for the other three types of heterogeneity there exist also differences regarding complexity, it is not sufficient to reduce the observation of inhomogeneity between taxonomies on merely the four categories of heterogeneity only. This complexity of heterogeneity depends on the used semantic, syntax, and fine granularity, to express one domain.

To accurately differentiate between the different types of heterogeneity existing by focusing on the different levels of complexity for the single four types, this chapter is presented. In the proposed methodology, we investigate the reasons that cause the four categories of taxonomic heterogeneity (linguistic, semantics, and data models) in a descriptive, and also for non-experts interpretable way. Hence, we propose to exploit different degrees of complexity (heterogeneity), in addition to the conventional four types of heterogeneity. The new set of degrees offers manifold benefits to help assessing the differences between two taxonomies and between background knowledge. Firstly, it can represent the characteristics of the existing taxonomy matching system in a discernible manner, instead of referring to only the standard metrics used in existing evaluation attempts, which are *F-Measure* (harmonic mean of precision and recall), *Precision* (retrieved and relevant instances), and *Recall* (relevant and retrieved instances) [53]. Through this, it can be more clearly stated, which concrete matching techniques and resources of background knowledge are available to overcome a specific degree of heterogeneity. However, in accordance with the results provided through existing evaluation attempts, e.g., held by OAEI, the degrees offer the possibility to justify why a system shows better results than other systems. Through this, it can be identified, which techniques and sources should be combined to achieve the highest matching accuracy. Finally, it helps highlighting open directions for the paradigm of taxonomy matching through assessing the heterogeneity between two taxonomies and sources of background knowledge more detailed. For example, which alternative sources of background knowledge could be used, or improvements regarding user involvement. The framework is directly applied on recent taxonomy matching systems and compared with results provided through the Ontology Alignment Evaluation Initiative (OAEI). The comparison highlights the effectiveness of the proposed method to identify capabilities of taxonomy/ontology matching systems and enables to identify open issues in the area of the research paradigm taxonomy matching.

The remainder of the chapter at hand is organized as follows. In Sect. 2.1, the methodology to differ the complexity for each type of heterogeneity by using degrees is presented. Hereby, a problem statement is given, meaning an explanation, when a specific type of heterogeneity is existing. In addition, the used degrees are explained in detail. In Sect. 2.2, an investigation regarding the four existing types of heterogeneity and its different degrees are presented, based on the before-gone methodology. Hereby, for each type of heterogeneity (terminological, conceptual, syntactical, semiotical), different degrees of heterogeneity are discussed. Afterward, the proposed methodology is subsequently utilized to evaluate the most recent matching (according to OAEI) frameworks, and examined against the obtained results (Sect. 2.3). The chapter concludes in Sect. 2.4.

2.1 Methodology of Taxonomic Heterogeneity

A **taxonomy** (T) is a collection of objects with similar properties to represent domains [130, 142]. Such objects are clustered by the help of a set (C) including concepts [158], which are related via edges (E) inside a directed graph. A **sub concept** (e.g., C_M) is detailing a superordinated concept (e.g., C_N), whereas a **super concept** is generalizing the subordinated concept. The concept with no superordinated concept is named **root concept** (e.g., C_R), which represents the most general concept. Two concepts sharing the identical superordinated concept are called **sibling concept**, e.g., C_M and C_S in Fig. 2.1a.

Fig. 2.1 A simple sample taxonomy T (**a**), a source taxonomy T_A representing different types of vehicles (**b**), and a heterogeneous target taxonomy T_B also representing vehicles (**c**)

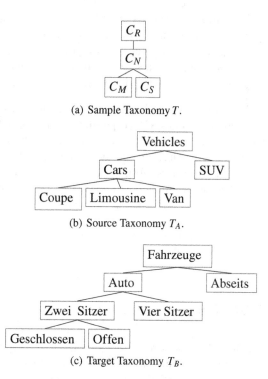

(a) Sample Taxonomy T.

(b) Source Taxonomy T_A.

(c) Target Taxonomy T_B.

Taxonomic heterogeneity is the disparity between two taxonomies describing the same domain of interest, for example when a British manufacturer of cars (source taxonomy T_A, see Fig. 2.1b) wants to order price lists from a German supplier producing sparkling plugs (target taxonomy T_B, see Fig. 2.1c), where $T_A \neq T_B$. The differences between both taxonomies concern the languages used to define the labels of the concepts, and the structures to model the relationships between those. To define such differences, [55] outlined four categories of heterogeneity:

- **Terminological heterogeneity** appears when the concepts of two taxonomies are expressed using different languages, e.g., T_A in English, and T_B in German.
- **Conceptual heterogeneity** arises if two taxonomies use different relationships to describe an identical domain. For example, "Cars" has three sub concepts ("Coupe", "Limousine", and "Van"), whereas "Auto" utilizes only two sub concepts to describe the same semantic fact ("Zwei Sitzer", and "Vier Sitzer").
- **Syntactical heterogeneity** occurs when for the storage of the taxonomies different data languages/models are used. For example, T_A is stored in *Extensible Markup Language* (XML) using a hierarchical manner to structure elements, where each category is assigned with a detailed description (e.g., "A coupe is a sportive car with three doors"), but T_B is captured inside a relational database, using formally described tables, and are only capturing the horizontal relationships (e.g., Name = "Zwei Sitzer", Parent = "Auto").
- **Semiotic heterogeneity** emerges when persons misinterpret concepts, respectively the relationships inside the taxonomy along with the used labels for concepts. For example, a customer does not expect a "Van" and a "Zwei Sitzer" in the same category (e.g., if "Cars" = "Auto"), because the one is used for transport, and the other for driving in a sporty way.

Matching taxonomies is the task of finding correspondences between two heterogeneous taxonomies. **Correspondences** are computed with a match operator [132], and stored in a four-tuple consisting of an identifier, a concept of the source taxonomy, a concept of the target taxonomy, and the theoretical relation between both concepts. The relationship uses subsumption ($\subsetneq$, $\supsetneq$), equivalence ($\equiv$), or disjointness ($\sqcup$) [14], to express the resulting correspondences. **Taxonomy Mapping** denotes the task of finding a set of such correspondences between two different taxonomies. Recent approaches add a truth-value in the form of a weighted boolean (usually between 0 and 1) to the correspondences, which state how sure the theoretical relation between concepts is [14, 93, 131].

2.1.1   Problem Statement

When reading the given notions for the four categories of taxonomic heterogeneity in common with the formulation to map two taxonomies, it is obvious that the heterogeneity substantially affects finding correspondences through a match operator.

Moreover, it can be claimed that the existing differentiation into merely four types of heterogeneity is not sufficient enough to compare matching techniques. This is because for each of the above-mentioned category, a more precise subdivision can be made, which has decisive impact on the matching quality result, e.g., resulted in one of OAEI campaigns.

For example, the manufacturer outlined in our sample has now two subsidiary companies. The already-mentioned supplier is located in Germany expressing the taxonomy inside a relational database. Another supplier is placed in China expressing its taxonomy with the *Web Ontology Language* (OWL).[1] As more translators and lexicons exist for treating English against German (and vice versa), and both share the same language family (Indo-European) and alphabet (Latin), the proposed techniques inside the taxonomy matching frameworks can more easily find certain correspondences between the taxonomies expressed in German and English, compared to the taxonomies expressed in Chinese and English, whereas in the literature both challenges are summarized as terminological heterogeneity. Furthermore, in the case of syntactical heterogeneity, as OWL is designed to represent rich and complex relationships between concepts by the help of markups, it is easier to find certain correspondences than inside the relational database, which is often just storing the super/sub concept relationships.

2.1.2 Degrees of Heterogeneity

Thus, for each of the four different categories of heterogeneity, various distinct degrees of heterogeneity can be distinguished. Such degrees characterize if a match operator can more easily or more difficult detect correspondences between the concepts inside two taxonomies, depending on the diversity of the used semantics, syntax, and fine granularity: **first degree** (lower heterogeneity), **second degree** (higher heterogeneity), and **third degree** (highest heterogeneity).

2.2 Heterogeneity Investigation

In the following, a classification method is proposed, which proves that taxonomy matching approaches are finding correspondences between two taxonomies harder or easier based on the complexity of heterogeneity they solve.

---

[1]http://www.w3.org/TR/owl-features/.

2.2.1 Terminological Heterogeneity

Terminological heterogeneity appears when concepts between two taxonomies are expressed in different languages, e.g., the source taxonomy is in English, in the following shortened as Θ_A, the target taxonomy in German, in the following shortened as Θ_B, and another target taxonomies in Chinese, in the following shortened as Θ_C. The different degrees for terminological heterogeneity depend on the accordance of the used alphabets and the semantics of the typology, i.e., interpretation of symbols, for the used characters:

- Low heterogeneity exists, if concepts are expressed in another thesaurus, e.g., Θ_A in English and Θ_B in German. As both languages merely differ in the words, but can be expressed throughout the same character encoding, which is used to represent a repertoire of characters, and many resources, e.g., dictionaries, exist to translate between languages. For example, a search for "English German Translation" resulted in 181,000,000 hits, the same query expressed in German resulted in 37,200,000 results, and it leads to the first degree of terminological heterogeneity.
- When using another alphabet to describe identical concepts, e.g., Θ_C in Chinese, the heterogeneity is higher. As now, words and characters are heterogeneous and the number of resources that can be used for translation as background knowledge are smaller. For example, a search for "English Chinese Translation" resulted in 108,000,000 hits, the same query expressed in Chinese resulted in 11,000,000 hits, and this disparity results to the second degree of terminological heterogeneity.
- Taxonomies, which differ in the kind of typology they are expressed, lead to the third degree. For example, if the target taxonomy utilizes an arbitrary abbreviation to express concepts, e.g., a concept carries the label "CA". As correspondences can only be measured with exploiting further comments assigned to the labels of concepts, e.g., using the description "represents a four-wheeled vehicle", which is of course not available for every taxonomy, this degree states the highest heterogeneity, namely the third degree of terminological heterogeneity.

2.2.2 Conceptual Heterogeneity

Conceptual heterogeneity arises if taxonomies are using different relationships to express an identical domain. For example, Θ_A utilizes three sub concepts to describe the super concept "Cars", whereas Θ_V utilizes only two sub concepts to describe the German translation of "Cars", namely "Auto", but details the sub concept "Zwei Sitzer" through two further less generalized concepts, namely "Geschlossen", and "Offen", to distinguish between "Coupes" and "Cabriolets". Thus, the diversity of generalization affects the degree of conceptual heterogeneity:

- When taxonomies differ in the number of levels, also known as ranks, but the number of concepts is identical, this leads to the first degree of heterogeneity. The reason is that the domain is describing the same artifacts, and all concepts of the source taxonomy can still be mapped to the target taxonomy.

- If two taxonomies contrast in the degree of generalization, e.g., Θ_A uses three sub concepts to describe the super concept "Cars", whereas Θ_B utilizes two sub concepts to describe the identical super concept. This leads to the second degree of conceptual heterogeneity. The relationship has to be measured by utilizing background knowledge because not all concepts can be mapped.
- If combining the both-mentioned degrees, it results to the third degree of conceptual heterogeneity. For example, if the source taxonomy includes a new super concept "Business" and utilizes two sub concepts for less generalization, but the target taxonomy features no "Business" concept. A complete mapping for all concepts is not possible.

2.2.3   Syntactical Heterogeneity

Syntactical heterogeneity occurs when for the storage of two taxonomies various data languages/models is used. For example, Θ_C is expressed in OWL and Θ_A in RDF. Both languages offer a set of markup languages to describe the semantic correlations between concepts. In contrast, Θ_B is expressed inside tuples of a relational database, not designed to describe the hierarchy of a taxonomy. Thus, the degree of syntactical heterogeneity depends on the variety between the used data languages, along with the possibility of the languages/models to express the semantic relationships between concepts:

- If two taxonomies are captured in different data languages, which are designed to describe taxonomies/ontologies, e.g., Θ_A in RDF and Θ_C in OWL. As OWL has a precise mapping to an RDF graph, thus is comparable with arbitrary RDF graphs, the lowest heterogeneity exists, namely the first degree of syntactical heterogeneity. These objects can be serialized in different syntaxes (e.g., **T**erse **RDF T**riple **L**anguage (Turtle) or RDF), yet be semantically equal.
- If two taxonomies are captured in different data languages using different markups to define attributes, e.g., the source taxonomy in OWL and the target taxonomy in LaTeX [94]. As both taxonomies utilize a hierarchical data model, but LaTeX does not offer predefined markup to define semantic relationships between concepts, the variety leads to the second degree of syntactical heterogeneity.
- In the case of capturing taxonomies in different data models, the highest heterogeneity exists, namely the third degree. For example, Θ_A is captured in RDF, but Θ_B uses a relational database. As both taxonomies are utilizing different models, both schemata have to be matched, before matching the concepts. Furthermore, as the relational database does not offer predefined attributes to define the semantic relationships between concepts, it complicates the matching.

2.2.4 Semiotic Heterogeneity

Semiotic heterogeneity emerges when persons disagree with the resulted matching, or the structure and labels used to describe the domain. As semiotic heterogeneity can be addressed by involving the expert in the matching task, the degree of semiotic heterogeneity differs depending on the enabled user involvement of the matching strategy:

- If the user is not satisfied with the mapping between concepts, but can correct the resulted correspondence, the matching algorithm is capable of supporting the first degree of semiotic heterogeneity. For example, "Sedan" and "Van" in the source taxonomy should be sub concepts of "Cars" and not sub concepts of "Family" in the target taxonomy. Thus, the user can delete the predicted correspondence.
- If the user can modify a resulted correspondence, for example through changing the theoretical relation between concepts, the matching approach is capable of supporting the second degree of semiotic heterogeneity. In the example above, the user can then decrease the value of subsumption between "Sedan" and "Family".
- If the user can create new mappings, the framework is capable of supporting the third degree of semiotic heterogeneity. In the example above, the user can then create a mapping between "Sedan" and "Van" with "Cars".

2.3 Experimental Evaluation

To examine the efficiency of the proposed methodology, we apply our approach on recent taxonomy matching systems. The three best performing systems, according to the latest OAEI campaigns, are considered for the experimental evaluation, which is (Not) **Y**et **A**nother **M**atcher ++(−) (YAM++(−)) [122], **A**greement **M**aker **L**ight (AML) [56], and **Log**ic-based and Scalable Ontology Matching/**Map**ping (LogMap) [83]. The experiment identifies, which degree for every type of heterogeneity is overcome by each approach. The review is summarized in Table 2.1.

When combining the results according to the achieved degrees for the different systems with the latest results achieved through the related OAEI campaigns, three types of determinations can be made. Firstly, for which type of heterogeneity the

Table 2.1 Degree of taxonomic heterogeneity comparison for different matching systems

Matching system	Year	Terminological	Conceptual	Syntactical	Semiotic
YAM++(−)	2013	2	3	2	3
AML	2014	2	3	2	2
LogMap	2014	2	3	2	1

taxonomy matching system perform well or badly? Secondly, why do some systems perform better or weaker for specific OAEI problems? And thirdly, for which type of heterogeneity the research area of taxonomy matching and semantic similarity assessment offers the most promising future directions? The results of terminological heterogeneity are compared to the multifarm track, where the approaches are aiming to solve the problem of multilingualism. Conceptual heterogeneity is matched to the library test case, a special track for matching conceptual heterogeneous taxonomies. Finally, the problem of semiotic heterogeneity is related to the interactive matching track, in which user involvement is investigated. Unfortunately, no OAEI track exists to evaluate the overcoming of syntactical heterogeneity.

To overcome the second degree of terminological heterogeneity, all three taxonomy matching systems are integrating a translating machine, mainly Google translator, or Microsoft Bing translator. LogMap further recalls Institute of Computing Technology Chinese Lexical Analysis System (ICTCLAS), to support the translation between English and Chinese vocabulary. As the corresponding OAEI campaign does not consider the problem when matching Chinese taxonomies, we can state that the frameworks utilizing the Microsoft Bing translator are performing much better to overcome terminological heterogeneity. As no framework attends to the third degree of terminological heterogeneity, future research can be expected for treating taxonomies expressed in an artificial language.

All frameworks are overcoming the highest grade of conceptual heterogeneity through the included matching techniques. According to the related OAEI campaign, the results fundamentally depend on the integrated background knowledge resources the frameworks are using. AML which showed the best matching quality results uses four sources. In contrast, LogMap only uses two sources. The number of background knowledge sources used are crucial for improving quality of matching results, as those enable to extract relations between concepts [157]. New resources of background knowledge that could be used are MULTIWordNet and GermaNet [68, 134]. As the named semantic lexicons are supporting non-English languages (German and Italian), those could help to overcome conceptual as well as terminological heterogeneity.

To overcome the second degree of syntactical heterogeneity, the three considered frameworks utilize the OWL API. It can parse the formats RDF, OWL, Turtle, Key Registration Service Specification (KRSS), and Open Biological and medical Ontologies (OBO). No framework can parse content of relational databases, and therefore, constitutes an open challenge for the field. First approaches that deal with this grade of heterogeneity are introduced in [154], or [161], for mapping relational databases to a RDF.

All three frameworks provide user involvement to reduce semiotic heterogeneity. Whereas LogMap and AML provide the involvement in the form of feedback and modification, YAM++(−) offers the user to create new correspondences. According to OAEI results, both evaluated frameworks could increase the matching quality. The framework providing the most user involvement, namely YAM++(−) did not participate in this track. Future research to increase the matching quality results can be expected for this type of heterogeneity by improving the involvement of the user.

2.4 Conclusions

In this work, we presented a new method to evaluate taxonomy matching approaches in an applied way, namely by reducing of taxonomic heterogeneity (terminological, conceptual, syntactical, and semiotic) and considering different degrees for each of the four categories of heterogeneity. The proposed method has been applied to the recently best performing taxonomy matching frameworks. The experimental evaluation has proven that the technique to consider the degrees of heterogeneity helps to support assessing the different taxonomy matching frameworks. In addition, the proposed method offers the possibility to highlight why frameworks are performing better or weaker in OAEI campaigns. The provided model further helps to justify three open questions in the field of taxonomy matching. Firstly, techniques to match taxonomies expressed in an artificial language are not existing. Secondly, the matching of taxonomies captured in different data models is not covered sufficiently. Thirdly, expanding the framework to provide more user involvement during the matching operation could improve the matching results.

The proposed methodology to evaluate taxonomy matching strategies, respectively the heterogeneity existing between taxonomies, has been applied to the recently best performing taxonomy matching systems. The experimental evaluation has proven that the technique to consider the degrees of heterogeneity helps to support assessing the different taxonomy matching frameworks, as well as to assess heterogeneity when matching with taxonomies acting as background knowledge. In addition, the proposed method has offered the possibility to highlight why frameworks are performing better or weaker in OAEI campaigns. The provided model further has helped to justify three open questions in the field of taxonomy matching, by identifying the until now not addressed degree for the different types of heterogeneity. Firstly, techniques to match taxonomies expressed in an artificial language are not existing. Secondly, the matching of taxonomies captured in different data models is not covered sufficiently. Thirdly, expanding the framework to provide more user involvement during the matching operation could improve the matching results.

Part II
Recent Matching Techniques, Algorithms, Systems, Evaluations, and Datasets

Chapter 3
Matching Techniques, Algorithms, and Systems

Abstract During the last years, the research area of taxonomy matching has made a massive progress, and many works in this area have been introduced providing significant advances to the field. The most improving attempts could highly outperform existing approaches because of two areas of innovation. The first innovation is to provide a combination of matching techniques inside a flexible matching strategy. Based on the domain, the used matching techniques and parameters can now be chosen in a flexible manner providing of course higher matching accuracy when overcoming one or multiple type(s) of heterogeneity. The second innovation is to support the matching techniques by linking to various sources of so-called background knowledge. The matching techniques, algorithms, and systems provide better quality and efficiency the more data and knowledge about the domain is provided. Using external knowledge in the form of thesaurus, linked data, or Semantic Web technologies, matching quality and efficiency could have been further increased. However, as no available survey considers recently developed matching systems according to the implemented matching strategies and utilized sources of background knowledge, a review focussing on this two criteria is required. To fill this gap, a comprehensive review is provided using this chapter.

Taxonomies are a subcategory of ontologies, which are using hierarchically ordered concepts to model a field of interest in a formal way. While keyword-search is known as a quick solution for finding specific products, a hierarchical representation of a domain has its merits for navigation and for exploring similar items. For example, own Web-shops classify goods in a hierarchically manner to help users finding relevant products [142] and to support searching for similar items.

However, as most of the enterprises, like Amazon[1] or eBay,[2] are using their own taxonomy to model over hundreds of interrelated concepts, a manual comparison between two repositories would be a time-intensive and error-prone task. To (semi)-

---

[1]http://www.amazon.com.

[2]http://www.ebay.com.

© Springer International Publishing AG 2017

H. Angermann and N. Ramzan, *Taxonomy Matching Using Background Knowledge*, https://doi.org/10.1007/978-3-319-72209-2_3

automatically detect matches between two heterogeneity resources, a broad research community is treating the paradigm of *Taxonomy Matching*. The heterogeneity can be categorized into four categories according to [157]: terminological heterogeneity (different labels/languages), conceptual heterogeneity (contradictory structures), syntactical heterogeneity (varying data models), and semiotic heterogeneity (disparate cognitive interpretations). Because the type(s) of heterogeneity that exist between two ontologies decisively affect finding correspondences, recent matching systems are focusing on the combination of multiple techniques in one matching strategy. According to the results obtained during the latest *Ontology Alignment Evaluation Initiative* (OAEI)[3] campaigns (in 2011–2015), such approaches have proven a significant improvement when overcoming one or multiple type(s) of heterogeneity with one matching system. As latest surveys in this area are focusing on data linking [59], the future challenges and progress between 2007–2010 [157], and the research areas directed in the last ten years [126], no work is reviewing recent systems against the actually used strategies that are the core component for interpreting the data inside taxonomies semantically correctly. On the contrary, the chapter at hand provides a review, which reduces the existing approaches on the available and actually utilized techniques to focus on the data exchange between two formal repositories, the key for a meaningful data mapping in nowadays distributed environments.

In detail, this work aims at providing a comprehensive investigation of recent matching strategies. By the help of profound comparisons, experts as well as non-experts can distinguish the strategies according to their specific requirements. The review includes an in-depth analysis of element-level and structure-level techniques, a comprehensive comparison of the most promising matching systems focusing on one or multiple type(s) of heterogeneity, and a summary of the most relevant evaluation approaches including progress obtained during the last five years on different datasets.

The remainder is organized as follows. Section 3.1 investigates the different matching techniques being available to overcome different types of heterogeneity. Section 3.2 reviews recently introduced matching algorithms according to the utilized matching strategy for overcoming at least one specific type of heterogeneity. Section 3.3 analyzes matching systems, which aim to overcome multiple types of heterogeneity with a single approach. The chapter concludes in Sect. 3.4.

3.1 Matching Techniques

Matching Techniques are single techniques, which are used by matching algorithms and matching systems to quantify the correspondences. Recent systems and algorithms are combining multiple techniques to treat different types of heterogeneity during one matching process, and to finally increase the matching quality result. According to the literature, nine types of matching techniques can be distinguished

---

[3]http://oaei.ontologymatching.org/.

[55]: five element-level techniques that point out similarities using literal values and four structure-level techniques utilizing the is-a structure.

3.1.1 Element-Level Matching Techniques

Element-Level Techniques use the literal values of the concepts, and/or its properties, to measure semantic similarity. Five element-level techniques exist: formal-resource-based, informal-resource-based, string-based, language-based, and constraint-based.

3.1.1.1 Formal-Resource-Based Matching Techniques

Techniques based on formal resources are referring to highly structured background knowledge in order to perform the matching task. The formal resources can be upper-level taxonomies, which summarize different domains in one repository, or domain-specific taxonomies, i.e., standard taxonomies, or resources published as linked data.

- *Domain-Specific* taxonomies represent concepts belonging to part of the world [125], e.g., the North American Industry Classification System describes customer groups of different domains [13].
- *Upper-Level* taxonomies represent the same domain as the taxonomies, but in a more generalized way, e.g., the North American Product Classification System describes product groups in a generalized way [116].
- *Linked Data* is the paradigm to structure and publish data for the Semantic Web, e.g., the service Freebase publishes databases for describing diverse artifacts [21, 23].

3.1.1.2 Informal-Resource-Based Matching Techniques

Informal-Resource-Based Techniques also exploit structured background knowledge in directories or annotations. However, now the external directories are informal ones.

- *Directories* are informally structured indexes, e.g., the Internet Retailing[4] directory is an informal Web directory.
- *Annotations* are any types of resources having further descriptions.

---

[4]http://internetretailing.net/.

3.1.1.3 String-Based Matching Techniques

Techniques that measure correspondences based on the equality of characters are called *String-Based* or *Lexical-Based* Techniques [36]. Such techniques aim in finding homogeneity between the labels used to distinguish between concepts, and/or between its descriptions. In taxonomy matching, such techniques are very often used to overcome terminological heterogeneity. The similarity can be evaluated against three types of sequences and result in a distance matrix.

- *Name Similarity* means to measure the similarity of single words or multiwords. The most well-known name-similarity measures are as follows: Levenshtein distance, Euclidean distance, Hellinger distance, Hamming distance, shortest path measure, Lin distance, Wu-Palmer distance, and Naive Bayes classifier. The Levenshtein distance characterizes the minimum number of edits required to transform one string into the other [96]. Each edit is qualified by the kind of transformation necessary for each character: if a single character has to be inserted, deleted, or substituted. The Euclidean distance depicts the length of the connection necessary to combine one point in the Euclidean space with another point. Hereby, each character of a string is assigned to a point in the Euclidean space [15]. The Hellinger distance depicts an alternative to the Euclidean distance but uses a probability density [71]. Hamming distance describes the number of characters being different at the same index. Each substitution necessary to transform the initial character into the target character increases the distance result [100]. The shortest path measure is the length of the shortest path between two nodes [95, 141]. Similar to Hamming distance, the strings must have the same length. The Lin distance measures the probability that a string occurs inside a label [89]. Wu-Palmer classifies each label according to its depth inside a comparable text corpus [172]. Naive Bayes classifier measures the probabilities under which a document in a class occurs. It does not take into account the importance of the words [62].
- *Description Similarity* takes into account compound terms to be compared with another sequence. Some important description-similarity measures are as follows: Term Frequency–Inverse Document Frequency (TF–IDF), Jaccard distance, and Cosine similarity. TF–IDF states how important a word is [86]. The value increases by the number of occurrences in the document. The Jaccard distance represents the resemblance between two sets of strings. The maximum coefficient is one, if the two sets are identical (for details, see [160]). Cosine similarity considers the sequences as vectors [160].
- *Global Namespace* considers the similarity between two namespaces.

3.1.1.4 Language-Based Matching Techniques

Language-Based Techniques consider the equality of the words used to classify concepts and are thus often combined with string-based methods. Such techniques take into account sequences of text, which are broken into meaningful elements to be compared. The comparison is predominantly supported with the inclusion of background knowledge. Such resources help to analyze the context of the compared concepts. Six categories exist to distinguish between the different types of language-based methods:

- *Tokenization* breaks a stream of text into words, phrases, symbols, or other meaningful tokens. For example, the stream "what're" can be tokenized into "what" and "are". Some recent algorithms techniques do not break the stream into single words as single tokens, as often two words can belong together, e.g., the stream "This is a car" is broken into "This is" and "a car". This method is named N-Gram, whereby the N depicts the number of words assigned to a token.
- *Lemmatization* groups together the different inflected forms of a word so they can be analyzed as a single item, e.g., the word "better" has "good" as its lemma.
- *Morphology* analyzes the inner structure of a given language's smallest grammatical unit, e.g., the word "SUV" is an abbreviation for "Sport Utility Vehicle".
- *Elimination* reduces the tokens with any elements considered as superfluous, e.g., the stop-word "and", or quotation marks are often removed in recent strategies.
- *Lexicons* are used to translate between languages. The most widely used lexicons, also named translators, are the Microsoft Bing[5] translator, or the Google[6] translator.
- *Thesauri* are utilized to analyze semantic similarity between concepts, e.g., the lexical database WordNet [58, 114]. In WordNet, a synset, i.e., set of synonyms, is presented for every word sense with one or more different word(s), and a class type for each word, which can be an adjective, adverb, noun, or verb. As WordNet is also a taxonomy, it helps to deduce conceptual similarities by the help of semantic relationships. The most important relationships are the hypernym relationships, i.e., super concept, hyponym relationships, i.e., sub concept, and the sibling term relationships, i.e., sibling concept. Other relationships, which are not provided for all synsets, are antonym relationships, i.e., opposite meaning, and meronym relationships, i.e., part of.
- *Word Sense Disambiguation* is utilized to analyze the sense of the sentence in the recent context, i.e., the most important sequence or token of the stream. For example, for the stream "We are driving car", "driving" is the sense of the sentence.

3.1.1.5 Constraint-Based Matching Techniques

Constraint-Based Techniques analyze the internal structure of the taxonomy. With the help of other methods, this type of techniques can be used to overcome conceptual heterogeneity. For the identification of similarity, such methods are considering the criteria used to structure the out-tree. The relations between the variables are stated in the form of constraints to analyze the correspondences.

---

[5]http://www.bing.com/translator/.

[6]https://translate.google.com.

- **Type Similarity** of the attributes is considered because these elements describe the concepts of a domain. Thus, two concepts sharing the same types, for example, "Sedan" and "Limousine" are sharing the attributes "Color", "Luggage Space", and "Number of Doors", can be assumed to be semantically similar. Or on the other side, when the attributes are terminologically different but hold the same meaning, for example, "Varnish" and "Color", concepts using this attributes can be considered as semantically similar.
- **Key Properties** are used to describe the concepts belonging to a taxonomy, e.g., the concepts can be classified after the "Sizes" of the different "Cars", or the "Price" of the different "Cars". When the concepts inside the taxonomies are structured according to a corresponding point of view, the taxonomies can be assumed to be similar.

3.1.2   Structure-Level Matching Techniques

Structure-Level Techniques use the formal structure inside the taxonomy to compute similarity between concepts. Four structure-level techniques exist: taxonomy-based, graph-based, instance-based, and model-based.

3.1.2.1   Taxonomy-Based Matching Techniques

Taxonomy-Based Techniques are focusing on the specialization (sub concept) and generalization (super concept) relationships inside the taxonomy, also named **is-a** relationship, and are mainly used for overcoming conceptual heterogeneity. There is no further classification provided in the literature for these techniques.

- **Taxonomy Structure** means the structure of the two out-trees to be compared. The taxonomies can differ in the total number of concepts used, in the number of relationships (sub/super concepts) utilized, and in the number of sibling concepts for each super concept. The less different the structure of two concepts is, through the assigned sub concepts, as well as through the superordinated concepts, the more semantically similar both are.

3.1.2.2   Graph-Based Matching Techniques

Graph-Based Techniques consider the taxonomies as labeled graphs. In contrast to the techniques mentioned in the previous section, now also the sibling relationships are taken into account allowing to compare sets of sub concepts and the distance between paths. Four directions of graph-based methods can be distinguished:

- **Graph Homomorphism** measures the relationships between nodes having different structures. For example, the graph $G_1 = (\{1, 2, 3, 4\}, \{1, 2\}, \{1, 2\}, \{2, 3\},$

$\{2, 3\}, \{2, 4\}, \{3, 4\}$) has the same relationships as $G - 2 = (\{1, 2, 3, 4\}, \{1, 2\}, \{2, 3\}, \{3, 4\}, \{2, 4\})$.

- *Path Similarity* measures the similarity depending on the series of edges used to connect following nodes. For example, G_1 uses six edges, G_2 only four.
- *Children Similarity* stands for the number of outgoing edges of a node. For example, the edge *1* in the graph $G_3 = (\{1, 2, 3\}, \{1, 2\}, \{1, 3\})$ has two children, the edges *2* and *3*.
- *Leaves Similarity* compares the vertices, i.e., common point between two edges, inside a graph representing the out-tree. For example, the graph G_1 has five vertices, the graph G_2 only four.

3.1.2.3 Instance-Based Matching Techniques

Instance-Based Techniques indicate the similarity between concepts depending on the instances (e.g., products and sub concepts) assigned to concepts, or sub concepts assigned to super concepts. The similarity depends on the two sets to be compared, because similar concepts should have similar instances.

- *Data Analysis* and *Data Statistics* means to compare the sets of instances or properties assigned to a concept. For example "X1", "X2", and "X3" are assigned to the concept "Coupe", and "X1", "X2", and "X4" are assigned to the concept "Two Seaters", so there must be a series between the three instances. Consequently, such methods are supported by other techniques to derive the similarity of the sets, e.g., with the Cosine or Jaccard distance.

3.1.2.4 Model-Based Matching Techniques

A few approaches have taken into account the description logics for overcoming taxonomy heterogeneity as well as the satisfiability, named *Model-Based* Techniques.

- *Satisfiability Solvers* determine if there exists an interpretation satisfying a given boolean operator, which can be true or false.
- *Description Logics* reasoner is a family of formal knowledge representation languages. A reasoner is a technique that is able to infer logical consequences from a set of entities.

3.2 Matching Algorithms

An algorithm for taxonomy matching uses a matching strategy consisting of matching technique(s), to overcome heterogeneity. Recent algorithms differ in the utilized matching techniques to overcome terminological, conceptual, syntactical, or semiotic heterogeneity.

3.2.1 Terminological/Conceptual Heterogeneity Matching Algorithms

The majority of recent taxonomy matching algorithms focus on combining different matching techniques to overcome terminological as well as conceptual heterogeneity with one matching strategy. Algorithms that focus on terminological heterogeneity intend to find correspondences when concepts of two taxonomies are multilingual, ambiguous, or synonym. Some algorithms deal with the terminological heterogeneity between two versions of one taxonomy. Other algorithms focus on the terminological heterogeneity of taxonomies that come from different providers. The recent algorithms mainly utilize string-based, language-based, and constraint-based techniques. Algorithms that center on conceptual heterogeneity investigate the different models inside the taxonomies. Hereby, the recent algorithms mainly use two different techniques: the taxonomy-based technique which utilizes the is-a structure of the out-tree on the one side and graph-based techniques which utilize the instances/entities assigned to a superordinating concept on the other side. For both types, the comparison is additionally supported by utilizing background knowledge.

For overcoming terminological heterogeneity between taxonomies of two different providers, an algorithm was presented that deals with the problem when various names are used to label identical concepts [127]. The approach combines techniques performing on the strings, the taxonomy, the graph, and the instances. The former-mentioned techniques are used to derive a base classification result. Hereby, the descriptions of products are analyzed with Naive Bayes, to compute a probability distribution over all target concepts. The resulting probability distribution is utilized to assign concepts in the target taxonomy to the products that come from the providers source taxonomy. Hereby, the taxonomy is used to structure the product catalog. The is-a structure is used to compute correspondences according to the path similarity of the sub concepts detailing the root concepts. The paths are compared using Cosine similarity. The final step computes a vector between source and target taxonomy and degrades the classification of a product to a concept. Another algorithm deals with terminological heterogeneity arising when the taxonomy evolves over time [46]. The algorithm uses techniques that perform on the strings, the languages, and the constraints. They first defined four different kinds of change patterns: total copy (labels are identical), total transfer (singular/plural form instead), partial copy (new sequence), and partial transfer (labels are different). A transfer denotes the total movement of an attribute over two time units. A copy stands for the partly movement of one attribute over two time units. The algorithm runs in two phases. First of all, it searches for candidate attributes in a set of predefined attributes by utilizing Bi-Gram, along with a ranking function. The second phase has the purpose to identify occurrences for the change patterns. It compares the similarity value of instances at run-time with a given threshold, after removing stop-words from the descriptions. The extension to the algorithm to also overcome conceptual heterogeneity was introduced in [175]. It mainly uses instance-based techniques and attends to the challenge when matching between objects is not possible because of missing links between them. First of all, the algorithm groups all related concepts, before all pairs of relationships are matched with each other. The overlap score is computed with Jaccard distance. Finally, the detected correspondences are added to the imported taxonomy. Another lexical-based algorithm was introduced that uses category name processing with string- and language-based techniques to compute correspondences between two different taxonomies [121]. The main contribution, in contrast to previous work, is the aggregated path similarity score that adds taxonomy-based and graph-based methods [123, 128]. It computes a normalized parent matching distance between candidate and score path. To compare concepts between two taxonomies, the algorithm utilizes the lexical database *WordNet*. This background knowledge is utilized to compare hyponyms and results in a split term set with utilizing Levenshtein and Jaccard distance. Finally, the candidate path identification computes the similarities for every path with computing co-occurrence, i.e., neighbor similarity, and order consistency, i.e., no occurrence contradiction. A self-configuring schema matching algorithm, which can automatically construct and adapt a matching process for overcoming terminological heterogeneity, is capable to automatically select and combine multiple matchers [132]. The main focus of the self-configuring schema matcher is on matrix features. Those aim to identify schema properties and the quality of matching rules. To evaluate the similarity of properties, the algorithm utilizes existing approaches: NameMeaningfulness [98], NodeTokenRatio [131], and Structural Similarity [98]. These are combined in the form of cross-matching (combining results) and multimatching (comparing results). The matching rules are divided into patterns. These describe a part of a process graph, the actions, the relevance, and an optional check box for the user. Another schema matching algorithm treating terminological heterogeneity due to multilingualism is instead instance-based and utilizes information across different language editions of Wikipedia [146]. Hereby, a first entity matching is applied based on two Wikipedia input articles that are identical but differ in the used language. Hereby, the algorithm makes use of the already in Wikipedia existing cross-language links. Logically, its outcome is the matching between identical articles of Wikipedia in different languages. As next, a template matching is performed that utilizes infobox template mappings and analyzes co-occurrences in the language versions. Finally, the attribute values of the articles are matched to find correspondences based on its constraints.

3.2.2   Syntactical Heterogeneity Matching Algorithms

Syntactical heterogeneity occurs when for the storage of the taxonomies different data languages/models are used. On the one hand, recent approaches deal with the mapping between different XML-based semantic languages (e.g., OWL and RDF). On the other hand, attention is paid to transform relational databases into semantic languages, but also to match schemas of two databases of the same model. This is because most real-world applications (e.g., in e-commerce) use relational databases to store documents. Thus, many taxonomies are still expressed inside relational databases that do not support semantic languages based on hierarchical database schemas. For that reason, some researchers started to develop approaches matching between semantic languages and relational databases. The recent algorithms center on this interaction between two different data models.

An algorithm for a direct mapping from relational databases to an RDF graph makes use of OWL vocabulary for the mapping and follows the direct mapping standard [11, 154]. The matching process itself consists of different sets of Datalog Rules (i.e., logic programming language) [1]. They are used as an interlayer to translate schemas, and afterward instances from relational databases to RDF. The rules generate Internationalized Resource Identifiers for concepts and data type properties. Finally, the relational schemas are translated to produce the RDF presentation. Another algorithm also focuses on the transformation from relational databases to RDF, and vice versa [161]. Contrary to previous approaches where only interested attributes are extracted, their method extracts every attribute and transforms the complete schema [3, 20]. After generating a text file, which describes the attributes, the data is extracted through Structured Query Language (SQL). Another semiautomatic schema matching for relational schemata to OWL taxonomies divides the transformation into two components [136]: a graph for representing the two syntactical different schemata in an unified way and the matching algorithm itself. Each graph represents a set of vertices, a set of direct edges, a set of labels, and a set of labels for the edges. After constructing the graphs for the relational schema and the ontology schema, the matching is performed with an extended version of the Similarity Flooding algorithm, which creates for two graphs a mapping based on the nodes [112]. Hereby, the annotations inside the ontology are utilized, as well as different types of user feedback, which utilizes the users' confirmation of mapping suggestions of previous iterations. Another algorithm focuses on answering SPARQL queries over relational databases [30]. Before the query can be performed, the relational database is transformed into an RDF, more precisely, the included concepts and its properties. Afterward, the RDF graph can be queried by translating the queries into SQL. To do so, the algorithm to translate between the different models consists of two main layers. The input layer that is capable to input taxonomies in OWL by utilizing OWL Application Programming Interface (API), taxonomies expressed as tuples by utilizing **Java Database Connectivity** (JDBC) API, as well as SPARQL queries by utilizing Sesame API. Based on the different inputs, the core layer is answering the SPARQL queries. To do so, this layer rewrites the SPARQL queries over the virtual RDF representation into SQL queries based on the relational schema. Another approach to overcome data type heterogeneity focuses on the mapping between OWL and Jess, i.e., rule engine for Java [63, 91]. The main goal is to integrate various taxonomies into a global taxonomy with multiple points of view. These views represent a single local taxonomy. The main components are matchers, which transfer alignments as axioms to the global taxonomy. They further insert description logic rules to the existing predefined semantic rules. Additionally, they use specific axioms for semantic integration and context rules in Semantic Web Rule Language [78].

3.2.3   Semiotical Heterogeneity Matching Algorithms

Semiotic heterogeneity emerges when human persons misinterpret concepts because of terminological and conceptual reasons. On the one side, the misinterpretation can be caused through an inadequate label assigned to a concept, e.g., the concept is named "Coupe", which most people understand as a car with three doors but there are also sportive limousine with five doors included as instances ("Shooting Brakes"). On the other side, the misinterpretation can be caused because of incorrect relationships between the concepts, e.g., a "Pick Up" is assigned to be a (is-a) "Car", but most persons consider this type of vehicles inside "Transporters" with the siblings "Van" and "Bus". For both issues, the semiotic heterogeneity is induced either through an incorrect matching, or through the expert herself/himself, e.g., when the provider considers the domain in a different way as the customer does.

To help understanding the expert-generated taxonomies by non-experts, researchers started to combine informal structured concepts, called folksonomies, with the formal concepts inside a taxonomy. A folksonomy, is similar to a taxonomy a set of concepts, so-called tags, but without $rootof()$, $subof()$ and $superof()$ relationships. Every concept defined in a folksonomy is a sibling ($sibof()$) of the other concept. The tags can be defined through the provider, or through the customers. This tags can be any type of label, e.g., a complain combined with an object (e.g., "Bad Seats"), the name of the instance itself (e.g., "X3"), or another subjective assessment (e.g., "Elegant", "Sportive", "Spacious"). In nowadays e-commerce applications, the tags are used in tag clouds, a method to visualize a set of tags inside a cloud where the tags are emphasized according to their usage frequency (see Fig. 3.1a), or the tags are directly added to the instances (see Fig. 3.1b). This type of metadata has shown high acceptance by the users, as it is not limited to the experts interpretation of the domain, is lightweight, is human understandable, and offers the possibility to create interlinked networks. However, it lacks in a sufficient support when navigating through the entire product range of a Web-shop, and the keywords used by customers contain semantic ambiguities and synonyms [101]. The combination of both metadata paradigms offers the possibility to add informal tags to the taxonomy. Thus, the taxonomy is not limited to the expert-generated concepts, but also includes the tags the customers would use for the concept (e.g., a subjective assessment like "Sportive" as an alternative for "Coupe") along with the relationships inside the taxonomy (see Fig. 3.1(c)). For example, a "Limousine" and a "Coupe" are for most customers an "Elegant" car, but no customer considers a "Van" as "Elegant", so "Coupe" is more related to a "Limousine" than to a "Van". Thus, the taxonomy could be transformed to classify "Limousine" and "Van" inside $subof(Elegant)$, and "Van" inside $subof(Spacious)$.

An algorithm that measures the similarity of users is using the subjective folksonomies from the user, which are assigned to taxonomic concepts [119]. The algorithm extracts the assessments with stop-word removal, through using Parts of Speech tuples (i.e., the class type of a token) [31], and comparison with a predefined set. Subsequently, the taxonomy is created automatically by using WordNet. Their approach

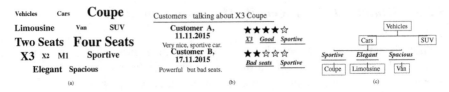

Fig. 3.1 A taxonomy including subjective assessments ("Elegant", "Spacious", "Sportive") represented as tag cloud (a), the user-created tags ("Good", "Bad Seats") and predefined tags ("X3", "Sportive") assigned to product reviews (b), and both types of metadata (folksonomy, taxonomy) combined inside one taxonomy (c)

is limited to English. The similarity measurement between the taxonomy and the folksonomy–taxonomy utilizes Cosine similarity. In contrast, another algorithm to overcome semiotic heterogeneity focuses more on the integration of preprocessed folksonomies into a static taxonomy [90]. The algorithm deals with the problem of various tagging from users in contrast to the expert-generated taxonomy. Resulting is a hybrid taxonomy–folksonomy set of concepts. Their algorithm uses existing data mining techniques: **Formal Concept Analysis** (FCA), i.e., to derive a taxonomy from a collection of objects, [171], and **Iterative Dichotomiser 3** (ID3) classification, i.e., container of metadata [140]. Formal Concept Analysis is applied to identify candidate tags and to reason about hierarchical relationships. Iterative Dichotomiser 3 is used to group the candidate tags, before the final integration into the taxonomy. A similar work for the direct recommendation of concepts to users utilizes SPARQL and combines content-based and knowledge-based filtering [38]. These filters are used to analyze the user's categorical and inner categorical preferences and to learn user's preferred weighting parameters. A content-based recommendation step processes a classifier on the base of user ratings on concepts. A knowledge-based recommendation step, in contrast, uses knowledge about how the concepts meet the user's needs. It finally processes a matcher between both. Another algorithm is based on existing classification approaches instead, but enhances the creation of taxonomies from folksonomies through creating tag pairs by the final users [6]. More precisely, they build up on the algorithm presented in [75], respectively its extension provided in [18]. Additionally, their extension requires the user to identify as much as they want is-a relationships between two tags, e.g., if "The Volkswagen Passat" is-a "Limousine" and/or a "Sedan". To create the final taxonomy from such tag pairs, the algorithm is affected by various parameters, which are occurrence threshold, similarity threshold, generality threshold, before creating the final taxonomy.

3.2.4 Analytical Summary Matching Algorithms

When comparing the algorithms focusing on terminological and conceptual heterogeneity, three meaningful conclusions can be drawn. Firstly, to overcome terminolog-

Table 3.1 Element-level techniques used to solve terminological heterogeneity

Algorithm	String-based	Language-based	Constraint-based
Papadimitriou et al. [127]	Naive Bayes	–	–
Nederstigt et al. [121]	Jaccard, Levenshtein	Tokenization, lemmatization, elimination, word sense disambiguation	–
Dinh et al. [46]	Bi-Gram, similarity threshold	Tokenization, elimination	Change patterns
Peukert et al. [132]	Repeating elements	NameMeaningfulness, NodeTokenRatio	Properties similarity

Table 3.2 Structure-level techniques used to solve conceptual heterogeneity

Algorithm	Taxonomy-based	Graph-based	Instance-based
Papadimitriou et al. [127]	Taxonomy structure	Path similarity, Cosine similarity	Product descriptions
Nederstigt et al. [121]	Parent matching	Path similarity, order consistency	–
Presutti et al. [175]	–	Overlap score	Class grouping, Jaccard
Peukert et al. [132]	–	Similarity matrix, CrossMatches	–

ical heterogeneity, most of the algorithms combine element-level techniques, namely string-based with language-based techniques (see Table 3.1). This combination has the advantage that not only the characters of the strings are compared, but also its context and the importance of a string in a sequence of text are taken into account. Furthermore, this combination provides the opportunity to remove non-indicative fragments like commas or vulgar languages. The recently applied string-based methods predominantly focus on the equality of the labels, as well as the similarity of the associated descriptions, e.g., with the Levenshtein distance. Secondly, with the help of WordNet, the most widely used lexicon to improve the matching quality, different semantic relationships can be detected, which cannot be derived from the is-a structure of the taxonomy, e.g., synonyms or antonyms. Recently, no strategy is capable when labels are expressed in an artificial language. Thirdly, to overcome conceptual heterogeneity, the algorithms predominantly add taxonomy-based, graph-based, or instance-based methods (see Table 3.2), which are considered as structure-level techniques. The two former-mentioned are focusing on the structure representing the taxonomy, either in the form of the is-a relationships or as a more complex graph. Hereby, the algorithms center on the relationships between sub- and super concepts to compute the distance between two concepts, or to infer the sibling concepts of the sub concept to be compared. Some approaches also apply instance-based methods.

Table 3.3 Element- and structure-level techniques used to solve syntactical heterogeneity

Technique	Constraint-based	Taxonomy-based	Graph-based	Model-based
Sequeda et al. [154]	OWL vocabulary	–	RDF schema	Datalog Rules
Thuy et al. [161]	Attribute description	Taxonomy structure	–	Description logics
Pinkel et al. [136]	–	Similarity flooding	Schema transformation, OWL annotations	–
Calvanese et al. [30]	–	–	Schema transformation, Sesame API, Java database connectivity, API, OWL API	–
Klai and Khadir [91]	Multiple points of view	–	Jess engine, Semantic Web rule language	Integrated distributed description logic

Those algorithms predominantly use Jaccard distance or Cosine similarity to infer the similarity based on the overlap score of two sets, respectively its descriptions.

Regarding syntactical heterogeneity, most of the reviewed algorithms focus on the combination of constraint-based, taxonomy-based, and model-based techniques to overcome disparity (see Table 3.3). One algorithm utilizes Datalog Rules to match between relational tables and RDF. Another similar approach is in contrast focusing on the is-a structure derived from the RDF graph and uses description logic to map attributes describing the taxonomy inside a hierarchical database against the properties describing a taxonomy in RDF.

Comparing the above studied algorithms aiming to overcome semiotic heterogeneity, it is evident that recent approaches started to combine folksonomies with taxonomies. These algorithms focus on the combination of language-based, taxonomy-based, graph-based, and instance-based techniques (see Table 3.4). The strategy either represents the folksonomy, respectively its concepts, also inside a graph for comparison with the taxonomy, or utilizes a third domain ontology. The language-based techniques are utilized to filter irrelevant tags, e.g., vulgarism, symbols, repeating elements, or not interpretable symbols. One algorithm is applying the TF–IDF measure to state the importance of a single tag. Two reviews focusing on the social classification of domains, and the combination of taxonomy and taxonomy are provided in [159, 177].

Matching systems are utilizing different matching techniques, respectively utilize a combination of multiple techniques, to computing a set of correspondences between two taxonomies. Some systems are treating multiple heterogeneities in one matching strategy, whereas other systems are focusing on a single heterogeneity. Thus, the systems are classified depending on the focus of the implemented strategy.

Table 3.4 Element- and structure-level techniques used to solve semiotic heterogeneity

Algorithm	String-based	Language-based	Taxonomy-based	Graph-based	Instance-based
Nakatsuji et al. [119]	–	Elimination, WordNet	Concept similarity	Folksonomy, Cosine	Sentiment analysis
Kiu et al. [90]	TF–IDF	Morphology, tags filter	Formal concept analysis	Folksonomy, Cosine	Iterative Dichotomiser 3
Cheng et al. [38]	–	–	SPARQL	–	Preference comparison, knowledge-based filtering
Almoqhim et. al [6]	–	Elimination	–	Tag pairs	–

On the one side, the most systems are focusing on terminological heterogeneity in common with conceptual heterogeneity with one strategy. On the other side, a few approaches are centering on a single heterogeneity, more precisely the matching between different data languages (syntactical heterogeneity), or are combining informal metadata (folksonomies) with taxonomies to overcome semiotic heterogeneity. The systems included in this section have all shown proper results during the latest OAEI campaigns, between 2011 and 2015, or have shown a significant improvement to the field. Furthermore, the included systems are either not considered in previous reviews provided by [55, 59, 126], or not analyzed in detail since the latest significant improvement. At the end of each type of heterogeneity, an analytical comparison is shown, allowing to draw multifaceted qualitative conclusions. Through this, it can be experienced, which combination of matching techniques is most commonly used to overcome a specific category of taxonomic heterogeneity, and, it can be derived, if new methods came to the forefront to improve this research area.

Matching systems are aiming to overcome multiple types of heterogeneity existing between two taxonomies. Hereby, the two formal repositories can be a source and a target taxonomy, or a target taxonomy and another taxonomy acting as background knowledge. During the last years, an increasing number of taxonomy matching systems has been proposed which prove a significant improvement according to the results shown in the latest OAEI campaigns (2011–2015). The OAEI is in the form of providing ontology/taxonomy specific tracks and test cases to evaluate taxonomy matching systems on the same basis for drawing conclusions regarding the best matching strategy. As it is held every year to provide a comparable evaluation regarding the improvement of existing and new matching systems, it is the most important matching evaluation in the field of taxonomy matching, but also in related fields like semantic similarity assessment. Through comparing the included matching strategies, it can be stated, which matching techniques should be utilized and combined to increase matching quality accuracy, which sources of background knowledge can be used to help assessing semantic similarity, if user involvement (i.e., involving the (non)-expert during the computation process) can increase accuracy, how the user involvement should be provided, etc.

3.3 Matching System

A system for matching taxonomies has the aim to overcome multiple types of taxonomic heterogeneity during one matching operation [126]. The reviewed systems have all participated at minimum one OAEI campaign in the last five years (2011–2015) [4, 34, 50, 52, 66], have showed proper results, and are either not compared in other surveys, or have not been compared since the latest significant enhancement.

3.3.1 Not yet Another Matcher (YAM++) Matching System

YAM$+ + (-)$ is a self-configuring, automatic, flexible, and extensible taxonomy and schema matching system [122]. It utilizes machine learning approaches, techniques derived from information retrieval, and background knowledge. The integrated information retrieval techniques aim to find a correspondence if no training data is given. The system is also able to match large-scale taxonomies by integrating a disk-based string method and utilizes WordNet as background knowledge. The system consists of eight components.

1. *Ontology Parser.* For parsing taxonomies, the system utilizes the OWL API [76], which can parse and write taxonomies in the formats RDF, OWL, and Turtle. It furthermore can parse the formats KRSS and OBO. In the case of multilingualism, the system utilizes Microsoft Bing translator, before normalizing with tokenization, removing stop-words, and stemming.
2. *Annotation Indexing and Structure Indexing.* To extract annotated information, e.g., labels or comments, the system applies the algorithm of Levenshtein, and N-Grams.
3. *Candidates Filtering.* The description filter is search-based and indexes entities through utilizing the Lucene Search Engine [108].
4. *Terminological Matcher.* Linguistic similarity is detected through applying the algorithms Lin, and Wu-Palmer distance.
5. *Instance-based Matcher.* The label filter is used to quickly detect candidate mappings based on instances. A label matcher splits the labels assigned to instances into tokens and calculates the information of each token for the complete taxonomy. A context profile matcher computes similarity values of entities with comparing the sets of instances assigned to a concept.
6. *Structural Matcher.* The structural matcher stores relationships between super- and sub concepts through applying two filters: the description filter and the label filter.

7. *Combination and Selection.* A dynamic weight factor aggregates the resulted mapping.
8. *Semantic Verification.* Finally, the semantic verification identifies inconsistent mappings with a Greedy Selection algorithm using Alcomol and Global Optimal Diagnosis methods [111]. Both have the aim to find the optimum result at each iteration. The user can additionally evaluate, modify, and add new mappings.

3.3.2   Agreement Maker Light (AML) Matching System

AML is an automated taxonomy matching system [56]. For lexical matching, it utilizes background knowledge. Its implementation is based on the predecessor AgreementMaker, which aimed to handle large taxonomies for the biomedical domain [42, 43]. Now, the recent system was changed to a general-purpose matching system. The system further provides full automation, supports multilingual matching problems, provides a repair module, and integrates structural matchers. The fully automatic support can be switched to an interactive matching task. The workflow for the automatic matching runs in nine steps.

1. *Loading Heterogeneous Taxonomies.* For loading taxonomies, the system utilizes OWL API. After parsing, the system transforms the taxonomies into an own data structure consisting of lexicon details, a property list, and a relationship map.
2. *Translation.* In the case of multilingualism, the system utilizes Microsoft Bing translator.
3. *Baseline Matching.* A lexical matcher, which is based on string equivalence, derives a first baseline alignment.
4. *Background Knowledge Matching.* As background knowledge, the system makes use of the resources Uber Anatomy Language (Uberon) [117], the Human Disease Ontology (DOID) [152], Me Subject Headings (MeSH), and WordNet.
5. *Word and String Matching.* For word and string matching, the system utilizes two different methods: a matcher for single words and a matcher for multiwords.
6. *Structural Matching.* A neighbor similarity matcher propagates the ancestors and descendants of the matched concepts. Hereby, it uses the distance weight factor. It so focuses on concepts which are expressed at various levels.
7. *Property Matching.* The property list stores names, types, domains, and ranges. The relationship map stores relationships between properties of concepts.
8. *Selection.* Hereby, the user can choose between three different selection strategies: the strict strategy for mapping one concept to exactly one another, the permissive strategy, where concurrent mappings are allowed, and the hybrid strategy, where maximum two mappings are allowed.
9. *Repair.* At this step, the user can provide optional feedback for the resulted mapping.

3.3.3 Generic Ontology Matching (GOMMA) Matching System

Generic Ontology Matching and Mapping Management (GOMMA) is originally a comprehensive infrastructure to analyze the evolution of life science taxonomies. For the latest OAEI evaluation of GOMMA, in 2012, the system was changed to a general-purpose matching system. This was achieved through removing the structural matching component. The current version also includes a generic component for semantically matching. The focus of GOMMA further lies in matching large-scale taxonomies. The system consists of eight components.

1. *Loading two Heterogeneous Taxonomies.* At the initial phase, the system utilizes Simple API for XML (SAX)[7] to parse and load XML-based taxonomies.
2. *Preprocessing.* The obtained information is stored in text attributes. Furthermore, the MyMemory translation API[8] is integrated in this step. It has the scope to automatically translate non-English concepts. After translation, a normalization step is integrated in which delimiters and stop-words are removed.
3. *Blocking.* At the blocking phase, the system determines an initial mapping. Firstly, it identifies a set of subgraph roots. Secondly, it computes the correspondences for those.
4. *Direct Matching.* The system can choose between two strategies, the direct matching, or the indirect matching. The former utilizes internal taxonomy knowledge. The system combines three different matchers: a matcher for synonyms and names, the matcher for comments, and a matcher for instances.
5. *Indirect Matching.* The indirect matcher, in contrast, utilizes external background knowledge, but uses the same matching techniques.
6. *Aggregation.* At the postprocessing phase, a filter, in the form of a threshold, aggregates the determined results of different matching strategies.
7. *Selection and Consistency Checking.* This last step stands for the involvement of the user in the matching task. He/she has to evaluate the resulted mapping.

3.3.4 Logic-Based Matching/Mapping (LogMap) Matching System

LogMap is a scalable system for matching rich taxonomies (>10,000 concepts) [83]. At the OAEI 2014 campaign, the LogMap family consisted of three subvariants: LogMap-C includes one more matching step, LogMap-Bio utilizes BioPortal as additional background knowledge, and LogMap-Lt as lightweight version that only supports string matching. The implementation is based on the predecessors LogMap [82] and LogMap 2 [84]. The new functionalities are the support of terminological

---

[7]http://www.saxproject.org.
[8]http://mymemory.translated.net/.

heterogeneity, an extended algorithm to repair correspondences, and by providing more user involvement. The system is divided into five theoretical elements.

1. *Lexical Indexation.* For loading taxonomies, the system utilizes the OWL API. The overcoming of terminological heterogeneity is realized using Google translator. The systems can also interpret Chinese language with ICTCLAS [176].
2. *Logic-based Module Extraction.* The size of the taxonomies is reduced with modularization techniques, i.e., dividing the taxonomies into modules.
3. *Propositional Horn Reasoning.* The relevant modules are encoded using Horn propositional representation, i.e., describing them as Horn clauses.
4. *Axiom Reasoning and Greed Repair.* The axioms are indexed through utilizing an interval labeling schema [2]. This represents an optimized data structure for storing directed acyclic graphs to reduce the cost of answering taxonomic queries [40, 120].
5. *Semantic Indexation.* All versions offer real-time user interaction during the matching process. The user has to provide feedback for computed correspondences.

3.3.5 Ontology Matching Reasoner (OMReasoner) Matching System

Ontology Matching Reasoner (OM-Reasoner) is an extensible system for matching taxonomies [156]. Multiple matchers can be integrated and combined inside the system. The system handles the matching task in literal and semantic levels and utilizes WordNet as external dictionary and for the interpretation of description logic for taxonomical reasoning. The system further makes use of the OWL description logics [77]. The system performs in three main steps.

1. *Parsing.* For parsing and interpreting taxonomies, the system utilizes the Jena ontology API [107]. This can parse the formats OWL, RDF, and DARPA Agent Markup Language (DAML).
2. *Combination of Matchers.* The system integrates the matchers based on Jaccard distance and WordNet. The first mentioned is used to detect literal correspondences. The last mentioned is applied to derive semantic correspondence from the external dictionary.
3. *Reasoning.* Reasoning has the purpose to derive further semantic correspondences. The system employs the extended description logic Jena and integrates a filter in the form of a flexible threshold. The combination of multiple match results is realized with two strategies: weighted summarizing and maximum method.

3.3.6 Risk Minimization/Instance Matching (RiMOM) Matching System

Risk Minimization-based Ontology Mapping-Instance Matching (RiMOM-IM) is a taxonomy matching system [155]. The latest system is an extended version of the system presented in [98]. The focus now is on interactive large-scale instance matching. It further integrates two new techniques to improve time performance: a method for blocking, i.e., the selection of candidate instance pairs, and an interactive matcher to allow user involvement. To improve accuracy of instance matching, a novel similarity aggregation method, called weighted exponential function, is applied. The system consists of five modules.

1. *Initial Interactive Configuration.* Initially, the user can choose the use-case-specific required modules and parameters (e.g., threshold). Only OWL files are supported as input format. Non-English taxonomies are translated using Google translator.
2. *Candidate Pair Generation.* Special symbols and stop-words are removed, before applying TF–IDF.
3. *Matching Score Calculation.* To pick smaller sets of candidate pairs, instead of detecting all pairs, blocking is applied in the form of inverted indexing.
4. *Instance Alignment.* For computing similarity, the Sigmoid measure [115], i.e., logistic function with an S curve, and exponential aggregation are utilized [155]. The last mentioned technique is applied to overcome the problem when various instance pairs use various numbers of aligned predicates. The system furthermore generates a priority queue. Hereby, a score matcher, which only executes the alignment with the highest correspondence, is used.
5. *Validation.* This module is optional and requests user feedback for all matched pairs.

3.3.7 Extensible Mapping (XMapxx) Matching System

Extensible Mapping $++$ ($XMap++$) is a scalable taxonomy alignment tool [49], which utilizes WordNet as background knowledge for string-based, linguistic-based, and structural-based matching tasks. The system combines those strategies with applying weights to the resulted matrices. The implementation of the latest version (2014) is based on its predecessors XMapGen and XMapSig [47]. The current version provides several new and improved components to increase accuracy and performance. Using Cosine similarity, involving particular and parallel matching, integrating Microsoft Bing translator in the case of multilingualism, and interfacing with the Java WordNet Library as background knowledge [61]. The system consists of seven components.

1. *Parser.* OWL and RDF can be used as input.

2. *String Matcher.* The string matcher compares textual descriptions of the source and target taxonomies.
3. *Linguistic Matcher.* Strings describing the concepts are compared with the Java WordNet library.
4. *Structural Matcher.* The structural matcher creates alignments based on structural proximity. It also utilizes WordNet for comparison.
5. *Combining Multiple Similarity Matrices.* The results of the previous matchers (linguistic, structural) are combined.
6. *Aggregation Operator.* Because all matchers create varying similarity matrices, an aggregator operator is used for the combination and trimming [48]. Depending on the strategy, three different aggregators can be used: aggregation, selection, or combination.
7. *Filtering.* Finally, a filter, in the form of a threshold, creates the final alignments.

3.3.8   Analytical Summary Matching Systems

When comparing the recent matching systems (summarized in Tables 3.5, 3.6, 3.7 and 3.8), three types of determinations can be made for the four different types of heterogeneity. Firstly, for which type of heterogeneity the taxonomy matching systems perform well or badly? Secondly, why do some systems perform better or weaker for specific heterogeneity problems? And thirdly, for which type of heterogeneity the research area of taxonomy matching offers the most promising future directions?

Table 3.5 Analytical comparison of mapping systems regarding terminological heterogeneity

System	Terminological
YAM++(-)	Microsoft Bing translator, tokenization, stemming, stop-word removal, Lin, Wu-Palmer, WordNet
AML	Microsoft Bing translator, string equivalence, multiword matching, Uberon, DOID, MeSH
GOMMA	MyMemory API, delimiters removal, stop-word removal
LogMap	Google translator, ICTCLAS, BioPortal, inverted indexing, interval labeling
OMReasoner	Edit distance, WordNet
RiMOM-IM	Google translator, WordNet, TF–IDF, delimiters removal, stop-word removal
XMap++	Microsoft Bing translator, triplet similarity, N-Gram, tokenization, Java WordNet

Table 3.6 Analytical comparison of mapping systems regarding conceptual heterogeneity

System	Conceptual
YAM++(-)	Levenshtein, Q-Grams, alcomol, greedy selection, Lucene search engine, dynamic weight factor
AML	Similarity inheritance, properties similarity, neighbor similarity, distance weight
GOMMA	Comment matcher, synonym matcher, instance aggregation
LogMap	Horn reasoning, Dowling-Gallier, interval labeling
OMReasoner	Jena description logic, weighted summarizing
RiMOM-IM	Similarity propagation, SIGMOID, exponential aggregation, vector distance
XMap++	Properties similarity, restriction similarity, associated weights

Table 3.7 Analytical comparison of mapping systems regarding syntactical heterogeneity

System	Syntactical
YAM++(-)	OWL API: RDF, OWL, Turtle, KRSS, OBO
AML	OWL API: RDF, OWL, Turtle, KRSS, OBO
GOMMA	SAX API: XML
LogMap	OWL API: RDF, OWL, Turtle, KRSS, OBO
OMReasoner	Jena API: OWL, RDF, DAML
RiMOM-IM	OWL
XMap++	OWL, RDF, RDF-S

Table 3.8 Analytical comparison of mapping systems regarding semiotic heterogeneity

System	Semiotic
YAM++(-)	Evaluation, modification
AML	Feedback
GOMMA	Evaluation
LogMap	Feedback
OMReasoner	–
RiMOM-IM	Feedback
XMap++	–

3.4 Conclusions

This chapter has presented a confrontation of the recently proposed matching systems and algorithms against the available matching techniques. In contrast to similar works, the review at hand put particular importance to the core components of the matching systems and algorithms, namely by reducing the approaches to its implemented matching strategies. The different matching techniques, as well as the discussed matching approaches, have been continuously presented in an applied way by using concrete examples and including comprehensive discussions. Now, it can be stated which techniques are available to overcome heterogeneity, which techniques are actually used, which datasets are available for evaluation, and in which directions future work can be expected.

Nine different techniques are recently distinguished for detecting correspondences between two taxonomies: five element-level techniques (formal-resource-based, informal-resource-based, string-based, languages-based, and constraint-based), and four structure-level techniques (taxonomy-based, graph-based, instance-based, and model-based). Whereas the former-mentioned techniques point out similarities by exploiting the labels of the concepts, respectively its descriptions, the latter named methods take advantage of the formal structure inside the taxonomy. The textual correspondences are in every reviewed system detected with the use of string-based methods (e.g., the Levenshtein distance) that are creating a distance metric depending on the equality of characters. The meaning of the words is additionally discovered with language-based methods, which are splitting textual sequences in meaningful tokens, and infers their semantic relationships (e.g., sibling terms and hypernyms) with the help of the lexicon WordNet. For deriving formal congruity, recent attempts predominantly are using the structure-level techniques based on taxonomies and graphs. While the former named are reduced to analyze the is-a relationships, the latter are considering the taxonomy in the form of a complex graph. Some approaches additionally started to additionally take into account the instances assigned to the concepts, e.g., the products classified in a Web-shop category. Such approaches are comparing the different sets of products with the Cosine similarity measure.

The confrontation of the available strategies against the actually used techniques inside the matching systems and algorithms has proven the present trend to combine multiple matching techniques inside one matching strategy. However, the used combinations differentiate decisively depending on the type of heterogeneity prioritized in the matching process. The most recently published matching systems are focusing to overcome terminological and conceptual heterogeneity with one matching strategy. To solve terminological heterogeneity, the most systems are combining the above-mentioned string- and language-based techniques to analyze the labels of the concepts not only regarding their characters, but also to derive the context of the words. Conceptual heterogeneity is predominantly treated through combining taxonomy-based with graph-based techniques. Other matching systems, in contrast, are only centering on a single type of heterogeneity with the used matching strategy, namely either on syntactical, or on semiotic heterogeneity. The systems focusing on syntactical heterogeneity aim to map between relational and hierarchical structured databases, as the available semantic formats OWL and RDF are rest upon the syntax of XML, but the most content is still stored in SQL-based databases. Such systems are using constraint-based methods in combination with taxonomy-based or model-based techniques. A pioneering trend to overcome semiotic heterogeneity, more precisely the misinterpretation of the formally structured concepts, is the unification with formal structured metadata (folksonomies). Such approaches are using graph-based methods to represent the folksonomy inside a graph, which is afterward compared with the is-a structure inside the taxonomy.

To overcome terminological heterogeneity arising through multilingualism, six of seven systems integrate a translating machine. $XMap++$, YAM++ (−), and AML make use of the Microsoft Bing translator. LogMap and RiMOMIM utilize the Google translator, and GOMMA utilizes the MyMemory API. LogMap integrates a second interface, named ICTCLAS, to translate between English and Chinese vocabulary. Regarding a comparable evaluation with the OntoFarm dataset, we can state that the systems utilizing the Bing translator are performing better to overcome multilingualism. To overcome heterogeneity arising through ambiguity and synonyms, all leading systems are combining string- and description-based techniques to improve matching quality results. Recently, no system is capable to compare concepts labeled with arbitrary abbreviations; thus, future research can be expected for treating taxonomies that utilize minimal lexical-based information.

All of the systems make use of background knowledge to detect similarity between two conceptual heterogeneous taxonomies. However, the systems differ in the number of sources they use, which substantially affects the matching quality results regarding two comparable taxonomy datasets. AML, which showed the best matching quality results, uses four sources. In contrast, LogMap only uses two sources. The number of background sources used are crucial for improving quality of matching results, as those enable to extract relations between taxonomy/ontology entities [157]. New resources of background knowledge that could be used for future research are as follows: MULTIWordNet and GermaNet [72, 134]. As the named semantic lexicons are supporting non-English languages (German and Italian), those could help to further overcome conceptual, as well as terminological heterogeneity.

Three systems utilize the OWL API to be capable of parsing various semantic data languages. It can parse the formats RDF, OWL, Turtle, KRSS, and OBO. OM-Reasoner, GOMMA, and $XMap++$ can parse various XML-based formats. RiMOM-IM can only input OWL. No system can parse content of relational databases and, therefore, constitutes an open challenge for taxonomy matching systems. The first algorithms that deal with this grade of heterogeneity are mapping relational databases to a RDF graph, e.g., in [154, 161].

Five systems utilize user involvement to solve semiotic heterogeneity. Whereas LogMap and AML provide the involvement in the form of feedback and modification, YAM++ (−) offers the user to create new correspondences. According to the results performed on the OAEI conference dataset, both evaluated systems (AML, and LogMap) could increase the matching quality. Future research to increase the matching quality results can be expected for this type of heterogeneity by involving the expert users through utilizing folksonomies, as suggested in recent algorithms [6, 38, 90, 119].

Chapter 4
Matching Evaluations and Datasets

Abstract The use of heterogeneous taxonomies requires the need of (semi-) automatic information processing and the computation of match scores. Taxonomic heterogeneity occurs in four different categories: terminological heterogeneity (different labels and/or languages are used to describe the concepts), conceptual heterogeneity (contradictory models, including a varying number of hierarchies), syntactical heterogeneity (varying semantic languages used and different syntax), and semiotic heterogeneity (disparate cognitive interpretations and misunderstanding). During the last five years, a large number of matching systems have been proposed, aiming to overcome one or multiple types of taxonomic heterogeneity existing between two taxonomies. The latest best performing matching systems and algorithms now combine multiple matching techniques to ensure the detected alignments, exploit different sources of background knowledge to extract further relations between taxonomy entities, and provide so-called user involvement to correct the resulted correspondences. Based on an analysis of the latest Ontology Alignment Evaluation Initiative (OAEI) results presented for different tracks and test cases between 2011 and 2015, this chapter provides a comprehensive review of state-of-the-art methods and attempts, as well as recent techniques, and discusses open challenges to each of the taxonomic heterogeneity categories.

Matching evaluation attempts aim to assess the implemented matching strategy against comparable challenges. On the one side, there are benchmarks that center on a comparable evaluation of the systems or algorithms. As measuring the quality of various matching approaches is still a challenging task, these approaches aim to create uniform metrics to measure the matching quality and matching efficiency. On the other side, datasets are proposed to compare the different systems to identical problems, because the matching quality strongly depends on the input taxonomies.

The most well-known matching evaluation attempt in the field is held by OAEI [10]. OAEI is in the form of providing ontology-/taxonomy-specific tracks and test cases to evaluate taxonomy matching systems on the same basis for drawing conclusions regarding the best matching strategy. As it is held every year to provide a comparable evaluation regarding the improvement of existing and new matching attempts, it is the most important matching evaluation in the field of taxonomy matching. Through

© Springer International Publishing AG 2017

H. Angermann and N. Ramzan, *Taxonomy Matching Using Background Knowledge*, https://doi.org/10.1007/978-3-319-72209-2_4

comparing the included matching strategies it can be stated, which matching techniques should be utilized and combined to increase matching quality accuracy, which sources of background knowledge can be used to help assessing semantic similarity, if user involvement (i.e., involving the (non)-expert during the computation process) can increase accuracy, how the user involvement should be provided, etc. During the recent years, an increasing number of taxonomy/ontology matching approaches has been proposed which prove a significant improvement in the field according to the results shown in the latest OAEI campaigns (2011–2015) [4, 34, 50, 52, 66]. However, no recent work is considering the progress indicated through the most recent campaigns, as the latest investigation against OAEI results was introduced by [157], where the authors analyzed the campaigns till 2010. Because there is an obvious need to analyze the progress in the field with the help of the most comparable evaluation approach, the chapter at hand builds on the observations presented in the previous chapters and logically investigates the campaign results between 2011 and 2015.

The analysis is furthermore used to outline the reason for the progress, namely by focusing, in contrast to other analysis, on the implemented matching strategies. The matching strategies are reviewed according to the utilized matching techniques and are considered regarding to the type of heterogeneity the approaches aim to overcome. Doing this, it can be clearly identified, which concrete techniques are combined to overcome one specific type of heterogeneity, why a matching strategy shows better matching quality results as compared to another one, and it helps to precisely identify open directions in the field.

The remainder of the chapter is organized as follows. In Sect. 4.1, the methodology used by OAEI for assessing matching approaches is presented. This includes a comprehensive review concerning the different utilized tracks and test cases presented, and the datasets used. In Sect. 4.2, the results obtained by different approaches are analyzed in detail, by summarizing and discussing the scores for each track and test case. Afterward, alternative evaluation attempts are discussed, before the chapter concludes in Sect. 4.4.

4.1 OAEI Matching Evaluation Methodology

OAEI campaigns have the aim to provide a methodology for drawing conclusions regarding the best matching strategy. To do so, the methodology is in the form of providing various tracks and test cases to evaluate taxonomy matching systems on the same basis. To experience the methodology in detail, more precisely the methodologies used between 2011 and 2015, this section is presented. It is analyzing the provided tracks and test and is analyzing the available and utilized datasets. In addition, it points out how many matching systems have participated in analyzed campaigns.

4.1.1 OAEI Matching Evaluation Participants

During 2011–2015, 49 taxonomy matching systems have participated in one of the campaigns; see Table 4.1. However, only one system has participated in all five campaigns, and only ten have participated in more than two campaigns. Furthermore, not all systems participated in all tracks and test cases. So measuring a comprehensive progress is challenging.

4.1.2 OAEI Matching Evaluation Methodologies and Datasets

OAEI is in the form of providing ontology-/taxonomy-specific tracks to evaluate taxonomy matching systems. It offers different tracks and test cases to provide a comparable evaluation for taxonomy matching systems with comparing frameworks on the same basis for drawing conclusions regarding the best matching strategies. Those are evaluated with measures inspired from information retrieval, namely Precision, Recall, and the aggregation of both with F-Measure analysis [157]. New tracks and test cases were provided during the years to react on the progress made in the field, as well as different datasets; see Table 4.2. The tracks and the aim of each track are as follows:

- The *Ontology Track* provides the conference test case, which consists of 16 taxonomies describing the domain of organizing conferences. This track can be used to evaluate taxonomy matching strategies according to the overcoming of terminological and conceptual heterogeneity.
- The focus of the *Multifarm Track* is to confront the systems with multilingualism. The problem is divided into two subproblems: matching taxonomies of the same domain of interest and taxonomies of different domains. This track uses the *OntoFarm* dataset. It includes seven taxonomies describing conferences. These taxonomies were translated into eight different languages (French, German, Dutch, Portuguese, Spanish, Czech, Chinese, and Russian), based on the English version.
- The *Directories and Thesauri Track*, more precisely, its library test case, aims at providing a matching task between the *Thesaurus Sozial-wissenschaften* (TheSoz) and the *Standard Thesaurus Wirtschaft* (STW)[1] thesaurus. STW represents an economical science thesaurus, whereas TheSoz depicts a social science thesaurus.
- The goal of the *Interactive Matching Track* is to show if user interaction can improve matching quality. For this track, the *biblio* dataset is used. Until now, this track only considers the expert user. However, OAEI has started to treat this challenge with included error rates.

---

[1]http://zbw.eu/stw.

Table 4.1 Ontology Alignment Evaluation Initiative participations 2011–2015

Framework	2015	2014	2013	2012	2011	$\sum$	$\sum$ Top 10
AM(L)	✓	✓	✓	–	✓	4	9
AOT(L)	–	✓	–	–	–	1	–
AROMA	–	–	–	✓	✓	2	1
ASE	–	–	–	✓	–	1	–
AUTOMSv2	–	–	–	✓	–	1	1
CIDER(-CL)	–	–	✓	–	✓	2	1
CLONA	✓	–	–	–	–	1	2
CODI	–	–	–	✓	✓	2	2
COMMAND	✓	–	–	–	–	1	1
CroMatcher	✓	–	✓	–	–	2	2
CSA	–	–	–	–	✓	1	–
DKP-AOM (-Lite)	✓	–	–	–	–	1	1
EXONA	✓	–	–	–	–	1	–
GMap	✓	–	–	–	–	1	–
GOMMA	–	–	–	✓	–	1	2
HerTUDA	–	–	✓	✓	–	2	1
HotMatch	–	–	✓	✓	–	2	–
IAMA	–	–	✓	–	–	1	1
InsMT(+/L)	✓	✓	–	–	–	2	–
JarvisOM	✓	–	–	–	–	1	–
Lily(-LOM)	✓	–	✓	–	✓	3	1
LDOA	–	–	–	–	✓	1	–
LogMap-C/Bio/Lt	✓	✓	✓	✓	✓	5	7
LYAM++	✓	–	–	–	–	1	–
MaasMatch	–	✓	✓	✓	✓	4	–
Mamba	✓	–	–	–	–	1	2
MapEVO/ PSO/SSS	–	–	✓	✓	✓	3	2
MEDLEY	–	–	–	✓	–	1	–
OACAS	–	–	–	–	✓	1	–
ODGOMS	–	–	✓	–	–	1	3
OMReasoner	–	✓	–	✓	✓	3	–
OntoK	–	–	✓	✓	–	2	–
Optima	–	–	–	✓	✓	2	–
RiMOM(-IM)	✓	✓	✓	–	–	3	–
RSDLWB	✓	✓	–	–	–	2	–
SBUEI	–	–	–	✓	–	1	–
SemSim	–	–	–	✓	–	1	–

(continued)

Table 4.1 (continued)

Framework	2015	2014	2013	2012	2011	∑	∑ Top 10
Serimi	–	–	–	–	✓	1	–
ServO-Map (-Lt)/OMBI	✓	–	✓	✓	–	3	3
SLINT++	–	–	✓	–	–	1	–
STRIM	✓	–	–	–	–	1	–
StringsAuto	–	–	✓	–	–	1	–
SYNTHESIS	–	–	✓	–	–	1	–
TOAST	–	–	–	✓	–	1	–
WeSeE-Match	–	–	✓	✓	–	2	1
WikiMatch	–	–	✓	✓	–	2	–
XMap(++)	✓	✓	✓	–	–	3	5
YAM++(-)	–	–	✓	✓	✓	3	7
Zhishi.links	–	–	–	–	✓	1	–
TOTAL (49)	22	13	21	23	18	–	–

Table 4.2 Ontology Alignment Evaluation Initiative methodologies 2011–2015

Track	Test case	Dataset	Year				
			2011	2012	2013	2014	2015
Benchmark	Benchmark	Biblio	–	✓	✓	✓	✓
		Ekaw	✓	–	–	–	–
		Finance	✓	✓	–	–	–
		Commerce	–	✓	–	–	–
		Bioinformatics	–	✓	–	–	–
		Product design	–	✓	–	–	–
		Cose	–	–	–	✓	–
		Dog	–	–	–	✓	–
		Energy	–	–	–	–	✓
Ontology	Anatomy	Adult mouse	✓	✓	✓	✓	✓
	Conference	Conference	✓	✓	✓	✓	✓
	Large bio	Large bio	–	✓	✓	✓	✓
Multilingual	Multifarm	Conference	–	✓	✓	✓	✓
Directories & Thesauri	Library	TheSoz	–	✓	✓	✓	✓
Interactive matching	Interactive	Conference	–	–	✓	✓	✓
Ontology alignment for question answering	OA4QA	Conference	–	–	–	✓	✓

- Other test cases are the *Large Biomedical Ontologies* with its focus on very large biological taxonomies, *Ontology Alignment for Question Answering* used to evaluate the exploitation of correspondences, and the *Anatomy* test case with its focus on biomedical ontologies.

4.2 OAEI Matching Evaluation Results

As experienced in the previous section, a total of 49 different matching systems have participated during the campaigns presented in 2011–2015. This allows summarizing and discussing the results obtained for the different systems, as well as the scores achieved for the various matching attempts. This is presented using the section at hand based on the results presented by OAEI [4, 34, 50, 52, 66]. For each track and test case, the results are analyzed by pointing out the progress made during the years, and the best performing systems. An analytical summary is provided afterward.

4.2.1 Benchmark Track

The benchmark track is identifying the areas in which each system performs strong or weak [157]. The systems run on systematically generated test cases. In 2012–2015, the biblio dataset was provided, which represents a bibliography taxonomy inspired by BibTeX. It is comparing an OWL description logic taxonomy having 97 concepts.

The three best systems of the last five years regarding benchmarks are Lily (0.90), YAM++(−) (0.89), and CroMatcher (0.88) [153]; see Fig. 4.1. However, the initial scope of this track, namely to see for which areas the systems perform strong or weak, is limited, as only one dataset has been used continuously over the years. However, Lily performed best for this domain, namely for bibliographic taxonomies, also for the campaigns before 2011.

4.2.2 Ontology Track

The ontology track provides three real-world taxonomies to provide different domain-specific test cases: anatomy, conference, and large biomedical. The anatomy test case aims to confront matching systems with expressive and realistic taxonomies of the biomedical domain. It includes two datasets: the Adult Mouse anatomy dataset, consisting of 2,744 concepts, and the National Cancer Institute Thesaurus, consisting of 3,304 concepts [69, 118]. The conference test case includes 16 taxonomies to describe the domain of organizing conferences. The large biomedical ontology test

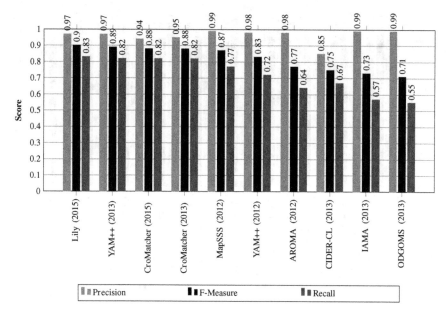

Fig. 4.1 Benchmark track. Comparison of matching quality results in 2012–2015

case has its focus on studying matching results for very large datasets (>10,000 concepts). This test case uses the largebio dataset.

As shown in Fig. 4.2, the three best systems, regarding F-Measure score, of the last four years for the anatomy test case are as follows: AML (0.94), AML-bk (0.94), and GOMMA-bk (0.92). Note that the abbreviation bk denotes a subversion of the systems utilizing various sources of background knowledge. The main observation of this test case highlights the improvement when including multiple resources of background knowledge and when the resources can be chosen dynamically. The system using merely WordNet, instead of including background knowledge of the biology domain, has a lower matching quality result accordingly.

For the conference test case, summarized in Fig. 4.3, YAM++(−) (0.71), AML 2015 (0.70), and Mamba (0.68) showed best results. During 2012–2015, the progress has slightly decreased, as previous versions of YAM++(−) show a 0.03 better F-Measure score than the best system of 2015 (Mamba) and a 0.04 better score than the best system of 2014 AML. The reason therefore compared to other test cases lies in the used resources of background knowledge. As leading systems mainly include resources derived from biology and biomedicine, no background knowledge is included focusing on conferences. For this reason, resources about conferences should be included in the future, as the predominantly used resource WordNet is too general. For example, the resource DBLP could be used [97].

YAM++(−) (0.82), AML (0.82), and XMap++-bk (0.82) are the three best systems of the last four years for the large biomedical ontology test case, as summarized in Fig. 4.4. The best system, which was presented in 2013, YAM++(−), expanded to

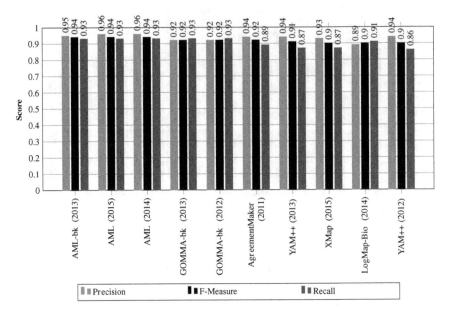

Fig. 4.2 Ontology track, anatomy test case. Comparison of matching quality results in 2012–2015

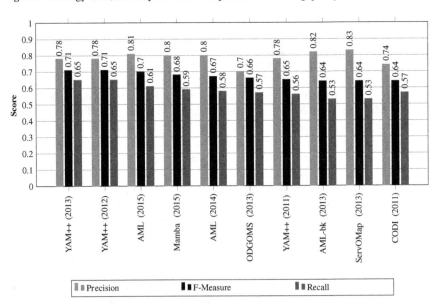

Fig. 4.3 Ontology track, conference test case. Comparison of matching quality results in 2012–2015

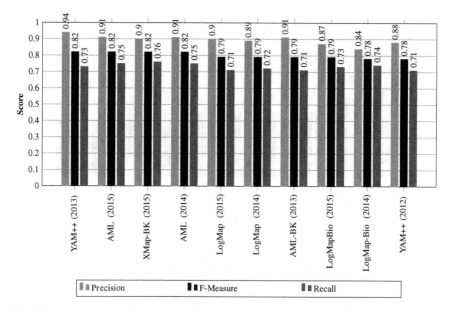

Fig. 4.4 Ontology track, large biomedical ontology test case. Comparison of matching quality results in 2012–2015

a large-scale matcher through utilizing a disk-based string method, through applying techniques from information retrieval for similarity measure, and through implementing a semantic verification method. This features helped to improve the quality by over +5.12%.

4.2.3 Multifarm Track

The focus of the multifarm track is to confront the systems with the case of multilingualism. The problem is divided into two subproblems: firstly, matching taxonomies of the same domain, and secondly, when dealing with taxonomies of different domains. The taxonomies are expressed in various languages (French, German, Dutch, Portuguese, Spanish, Czech, Chinese, and Russian), based on the English version of the OntoFarm dataset [110].

The three best performing systems for matching taxonomies of different domains are as follows: AML (0.54), LogMap (0.41), and WeSeE-Match [129] (0.41), shown in Fig. 4.5. In contrast, for the mapping of terminological heterogeneous taxonomies of the same domain, the three best performing systems are as follows: MapSSS (0.66) [32], AML (0.64), and Codi (0.62) [79], shown in Fig. 4.6. The core improvement of this track was achieved through utilizing a translating service, either Google or Microsoft Bing translator. However, compared to the other tracks, the multifarm track

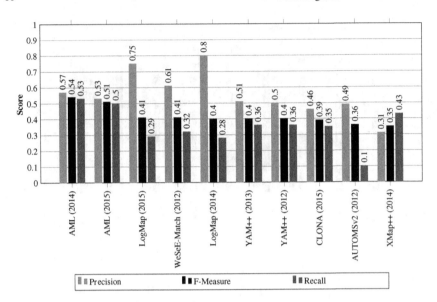

Fig. 4.5 Multifarm track, different ontologies. Comparison of matching quality results in 2012–2015

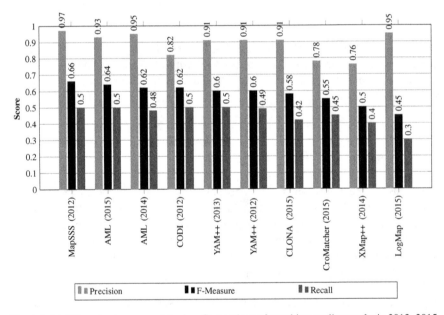

Fig. 4.6 Multifarm track, same ontologies. Comparison of matching quality results in 2012–2015

offers the most space for improvement. One possibility to improve the results for this track would be to include further information about the translations. For example, Google and Bing do only provide the word in another language along with a small explanation. However, more information about the words to be translated/compared, e.g., through using an informal directory in addition, would allow gathering more information about the semantic meaning of the concepts. Another problem is the language of the resources of background knowledge, which are also mainly in English. Resources in other languages could further help to overcome multilingualism; for example, an Italian version of WordNet exists, as well as a German version. The differences between matching taxonomies of different domains, and when matching terminological, heterogeneous taxonomies of same domains does not rely on the used translators, but on the used matching strategies.

4.2.4 Directories and Thesauri Track

The directories and thesauri track, more precisely, its library test case, aims at providing a matching task between the Thesaurus Sozialwissenschaften and the Standard Thesaurus Wirtschaft [64, 106]. The latter represents an economical science thesaurus, whereas the former depicts a social science thesaurus. The dataset consists of specific versions of those thesauri. A SKOS version, a version expressed in OWL, and an OWL version including SKOS labels.

The three best systems, as shown in Fig. 4.7, are as follows: AML (0.80), ODGOMS (0.76) [92], and YAM++(−) (0.74). One key observation is that since this track has been provided (2012–2014), a continuous improvement has been performed. This shows that the combination of various matching techniques inside one matching strategy also helps in improving matching quality, even if the repository is limited to the is-a structure.

4.2.5 Interactive Matching Track

The goal of this track is to show if user interaction can improve the matching quality result. Only 13 systems (including subversions) have participated in this track between 2013 and 2015. For this track, the dataset from the conference track is used without any adjustments.

The comparison of the obtained results evidenced that the increase in the F-Measure score depends significantly on the system itself, more precisely on the quality resulted when no user is involved during the matching task. The range of improvement of the ten best performing systems goes from +346.75 to 1.46%, as shown in Fig. 4.8, where the best performing system has shown very low quality when not involving the user (0.17) and its result is thus not meaningful. Furthermore, not all systems that have participated in this track could improve the matching

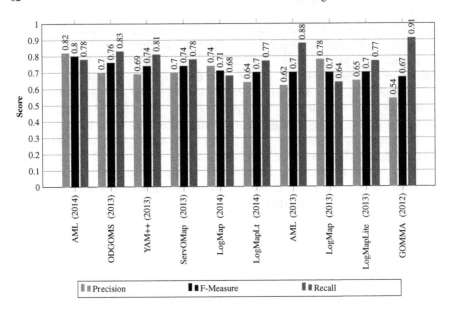

Fig. 4.7 Directories and thesauri track, library test case. Comparison of matching quality results in 2012–2015

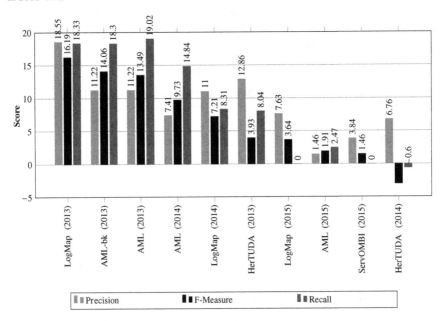

Fig. 4.8 Interactive matching track. Comparison of matching quality results in 2013–2015

quality result: HerTUDA [74] version 2013 (−3.00%), WeSeE-Match version 2013 (−14.00%), and WeSeE-Match version 2014 (−22.46%). These results demonstrate that in general user involvement can help in assessing semantic similarity; however, the type of user involvement is crucial. Whereas the leading system provides the user involvement in several directions (evaluation, modification, feedback), the other systems are only focusing on the evaluation itself. In addition, the leading systems can use the user involvement for training the system for the next matching operation; thus, the systems can learn from correct and incorrect matchings.

4.2.6 Analytical Summary

As can be seen from the results' comparison, the various systems participate not in all test cases, and also not during all campaigns. For this reason, measuring the progress of each system is challenging. Same for measuring the progress for the different test cases and tracks. However, a summary for the tracks possible to compare is presented in Fig. 4.9, which allows the following conclusions regarding the progress of the systems to overcome different types of heterogeneity, regarding the progress during OAEI campaigns, and finally, regarding future directions for OAEI.

During 2011–2015, the matching system AML, respectively AML-bk, LogMap, and YAM++(−), showed the best results on different OAEI tracks. Through this, the progress made in the field predominantly depends on the enhancement of the before-mentioned systems, which has four main reasons. Firstly, to match taxonomies of different languages, recent systems integrate a translator like Google or Microsoft Bing. In addition, the usage of translators allows to also apply background knowledge not expressed in English, even for non-English taxonomies. For example, the lexicon WordNet only contains English synsets, but with the help of translators, the labels of any language can be compared with WordNet to improve matching quality. Secondly, the recent systems are capable to input taxonomies that are stored in various XML-based semantic data languages. Thirdly, leading systems provide different possibilities to also involve the user during the matching operation, either through providing feedback, or through evaluating the obtained correspondences. For most of the systems, the user involvement could significantly help in increasing matching quality. Of course, the improvement significantly depends on the result when not involving the user. Fourthly, the systems utilize a variation of matching techniques and different resources of background knowledge in one matching strategy. Through this, the systems are capable to treat multiple types of heterogeneity with one system and do not overspecialize on a single domain. The aggregation used by all systems is to combine string-based with description-based techniques. This allows removing irrelevant text in the label or the description and allows analyzing single labels, as well as labels consisting of multiple sequences.

The progress for OAEI tracks has increased accordingly with the enhancement of the best performing systems. The usage of translators significantly helped the systems to extensively increase quality for the multifarm track, namely +24.39%. Since the

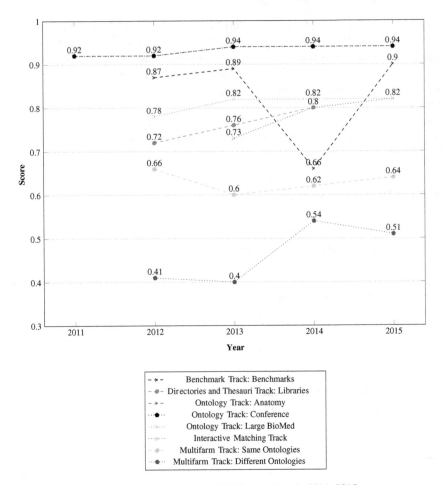

Fig. 4.9 Progress for the different test cases of OAEI campaigns in 2011–2015

interactive matching track was introduced in 2013 for the first time, and only 13 systems participated during the last three years, the conclusion is limited. The best system for this track is LogMap, which could increase the matching quality result by +16.19% when involving the user. This demonstrates that the user involvement in the form of feedback is most effective, namely when the feedback is taken to train the system for the next matching operation. The usage of various sources of background knowledge to increase matching quality showed the most impact when matching directories, i.e., repositories limited to the is-a structure. For this reason, the progress during 2012–2013 for the directories and thesauri track is +12.76%, and during 2013–2014, it is +5.26%. Compared to the other tracks, a crucial improvement of +13.89% can be measured during 2012–2015. When concluding the ontology track, the observation must usually be divided into the three specific test cases representing different domains (anatomy, conference, and large biomedical ontologies). However, during recent years the improvement for all three test cases was lower compared to other tracks, namely +2.17% for the anatomy as well as for the conference test cases, and +5.12% for the large biomedical ontology test case. In all three test cases, either AML or YAM++(−) performed best. An exception was provided in the year 2012, when a subversion of GOMMA including various sources of background knowledge showed the best results for the anatomy test case. Again, the improvement relies on the usage of various sources of background knowledge and the combination of different techniques. In contrast to the directories and thesauri track, the ontology track provides more specific test cases, and the test cases are based on the core domain of ontologies (biomedical). This is the reason for the lower increase compared to the other tracks, as already high results have been achieved before 2011. For the benchmark track, which has the aim to show for which area the systems perform strong or weak, only the biblio dataset was used continuously during 2012–2015. However, this track confirms a progress of +3.44%. Only in 2014, the systems have shown a considerable decline compared to the previous year.

As discussed previously, OAEI provided various sets of tracks, test cases, as well as datasets to react on the progress in the field of taxonomy and ontology matching. However, when comparing the types of heterogeneity the systems and algorithms aim to overcome, some new tracks would be required. Regarding terminological heterogeneity, recently no test case is considering taxonomies that use labels with a less semantic meaning, e.g., artificial language, or acronyms. However, a dataset providing this analysis could help to improve the systems' structure-level-based techniques, as the element-level techniques are limited. In addition, techniques could be expected that are capable to learn from the abbreviations and their similarity compared to the descriptions and properties. Regarding syntactical heterogeneity, no test case is considering the need to also parse and match taxonomies stored in relational databases, although many domains store the super and sub concept relationships inside relational schemas (e.g., in e-commerce). In contrast to previous years, where the schema matching approaches had to be evaluated manually regarding matching quality and performance, a considerable progress has been made during the last years [17]. For example, a schema matching benchmark named EMBench was introduced, which provides the generation of benchmark test cases to evaluate entity matching systems [81]. In contrast to the static datasets provided through OAEI, EMBench offers the possibility to generate the evaluation dataset with dynamic complexity, dynamic levels of types, as well as different levels of scales. Or, XBenchMatch was introduced that is also a schema matching benchmark, but is based on already available datasets [51]. The included datasets are also flexible to consider the different degrees of heterogeneity and complexity existing between the schemas. For example, three different types datasets are considered regarding terminological heterogeneity, the size of the schema is also classified in three categories of complexity, and two different sets of datasets consider the different grades of conceptual heterogeneity for schema matching. Regarding semiotic heterogeneity, the most lack of a comprehensive evaluation exists. Until now, the efficiency of user involvement to filter for irrelevant mappings is treated with error rates. However, the obtained results do not distinguish between expert and non-expert users, which of course distorts the results. In addition, the main progress regarding semiotic heterogeneity algorithms is not considered, namely when matching the taxonomy with informal repositories, i.e., folksonomies. This would allow assessing the system also if one repository does not provide any structure-level-based analysis possibilities.

4.3 Alternative Evaluation Attempts and Datasets

Besides OAEI, other attempts exist aiming to evaluate matching systems, algorithms, and techniques, as well as including the implemented matching strategies. Analogous to the tracks and test cases analyzed above, those works also help in assessing the available matching approaches. The works introduced recently include evaluation metrics, benchmarks, and novels datasets to evaluate the matching strategies. To experience such approaches, this section is used. For each approach, the core functionality and aim are presented in detail.

One recent benchmark approach provides a catalog of real-world data exchange patterns [147]. The patterns pay attention to terminological, syntactical, and conceptual heterogeneity. Furthermore, the authors determined two kinds of parameters for evaluating matchers: structure, which means the number of levels, and data parameters, which means related concepts and data properties that are standing for the number of individuals, data properties, and object properties. Another suite is focusing on providing conference ontologies derived from OAEI, and the created corresponding relational databases [137]. This benchmark aims at evaluating a wide range of mapping tasks in relation to ontology data integration. Recently, the suite consists of three conference ontologies provided through OAEI. Three different types of challenges are considered inside the suite: naming conflicts, structural heterogeneity, and semantic heterogeneity. The former considers the usage of different conventions to name the concepts either through using identifiers or abbreviations instead of technical artifacts, or through using synonyms (i.e., terminological heterogeneity). The structural/conceptual heterogeneity challenge includes differences because of the different modeling constructs (type conflicts), differences because of the representation inside the schema (key conflicts), and homogeneity because of different dependencies (i.e., 1:1, 1:N, N:N; dependency conflicts). The latter considers the differences because of an impedance mismatch, and semantic expressiveness. However, the most well-known campaign to evaluate taxonomy matching systems is OAEI. With its first campaign in 2005, OAEI provides different specific tracks to compare systems on the same basis for drawing conclusions regarding the best matching strategies. Recently, tracks for overcoming terminological heterogeneity, and for overcoming conceptual heterogeneity exist, but also a new track to evaluate the overcoming of semiotic heterogeneity through user involvement is included. The campaign is performed annually, and the participating systems are evaluated with measures inspired from information retrieval, namely Precision (measure of correctness), Recall (measure of completeness), and the aggregation of both with

F-Measure analysis. During 2011–2015, 49 different matching systems have participated in minimum one track; for further details, see Table 4.1. However, only one system has participated in all five campaigns, and only ten have participated in more than two campaigns. Furthermore, not all systems participated in all tracks and test cases. So measuring a comprehensive progress of each system is challenging. In addition, new tracks and test cases were provided during the years to react on the progress made in the field; see Table 4.2. For this reason, the following presented comparison is based on OAEI test cases that were provided in minimum three campaigns, where the identical datasets have been used to identify the quality of the different systems.

4.4 Conclusions

This chapter has evaluated the progress during the last five years in the field of taxonomy and ontology matching. This has been achieved using the results obtained in OAEI campaigns in 2011–2015. This has been provided to prove the progress of different matching systems and for different matching challenges. In addition, alternative evaluation approaches have been discussed as well.

The obtained results of the evaluation comparison have evidenced that the advancement in the field depends on the given matching task. The most significant progress has been made when matching taxonomies that are expressed in different natural languages. This progress is achieved by integrating a translating machine (e.g., Microsoft Bing translator or Google translator) during the matching task. Progress has also been made for matching directories. The main reason is that recent taxonomy matching systems are now combining different sources of background knowledge to infer the semantics between the taxonomies (e.g., in AML, YAM++(−), and GOMMA). The provided user involvement has also been improved by recent systems. Those provide user involvement in the form of requiring feedback from the user or are able to learn from through the user validated correspondences. The leading systems like AML, LogMap, and YAM++(−) differentiate from other approaches by utilizing various sources of background knowledge; e.g., the matching system AML uses four sources: Uberon, DOID, MeSH, and WordNet. Furthermore, the best systems according to OAEI results are combining different matching techniques inside one matching strategy, e.g., graph-based, instance-based or property-based.

The first direction for further works concerns taking into account background sources that are not expressed in English language, e.g., approaches that provide semantic lexicons in other languages like MULTIWordNet (Italian) or GermaNet (German). The second direction concerns the evaluation of matching syntactical heterogeneous taxonomies. Recent matching algorithms that aim to match syntactical heterogeneous taxonomies are focusing on a matching between taxonomies stored in relational databases and taxonomies stored in semantic languages. However, the comparison of OAEI methodologies has shown that there is recently no track provided to evaluate such approaches. Thirdly, as user involvement promises high potential to increase the matching quality results, it can be expected that this optional matching step will be improved. Until now, not all systems are capable to provide techniques for supporting the user when validating or correcting the detected correspondences.

Part III
Taxonomy Heterogeneity Applications

Chapter 5
Related Areas

Abstract Works on taxonomy matching have the aim to help overcoming taxonomic heterogeneity existing between two taxonomies. The heterogeneity arises, because different persons usually have a varying cognitive and methodological interpretation of a domain, even if the domain of interest is same. However, when using taxonomies to structure data, the heterogeneity inside taxonomies provides significant benefits for other research areas. The heterogeneity in the form of concepts existing can be used to help creating taxonomies that only filter for relevant concepts according to a provided keywords. The heterogeneity can also be utilized to create subtaxonomies based on the preferences shown by a user or customer. And, the heterogeneity can be used to adapt the taxonomy by providing different modification rules, also by utilizing preferences provided through a recommender system. Using the chapter at hand, all the mentioned research paradigms are discussed in detail, including the areas of: dynamic taxonomies, catalog segmentation, personalized directories as well as recommender systems. For each area, the most important techniques as well as the most recent applications are discussed, after detailing the aim of the research area.

During the last decade, the amount and usage of information has dramatically increased [9]. The retailing sector, for example, has expanded significantly [166]. Nowadays, the digital marketplaces provide an extensive number of products and services to their customers, and the commercialization is done using different online and offline marketing strategies, including the usage of various channels and devices to enter the mainly multichannel marketplaces [27].

Across the different marketplaces and channels, taxonomies play the crucial factor for structuring information and for finding the desired information at the right time and place. The reason is that in contrast to folksonomies, those are using hierarchically ordered concepts to model a field of interest in a formal way [67]. This hierarchical representation of a domain has consequently its merits for navigation and for exploring similar items [142]. However, as the taxonomy is created through the expert, or by matching with a standard taxonomy, taxonomies lack of acceptance, as every individual has his/her own cognitive embossing to categorize elements of an identical field. In addition, the various marketplaces and channels have different capabilities regarding the storage and visualization of an identical taxonomy. To

© Springer International Publishing AG 2017

H. Angermann and N. Ramzan, *Taxonomy Matching Using Background Knowledge*, https://doi.org/10.1007/978-3-319-72209-2_5

overcome this, different research paradigms have arisen, aiming to evolve the taxonomy according to specific requirements. Those are the research areas of: catalog segmentation, dynamic taxonomies, and personalized directories [9]. The works on catalog segmentation have the aim to reduce the taxonomy according to the concepts a customer actually requires. To do so, subtaxonomies (catalogs) showing only a subset of the concepts are predefined. Afterward, the customers are assigned to one subtaxonomy satisfying mostly the customers needs. Attempts in dynamic taxonomies, including the subarea of faceted search, adapt the taxonomy according to a search query performed through the customer. Hereby, the keyword is matched with the concepts of the taxonomy and the sub concepts of the matching concept are shown for further filtering for the actually desired product. Works on personalized directories try to adapt the taxonomy according to the hierarchical embosing of the customer. Latest works in this area propose adaption rules for transforming the taxonomy without reducing it semantically. All of the above-mentioned areas are usually supported by a fourth area, namely recommender systems. Works in this area have the aim to experience the customers' requirements by analyzing products purchased, queries performed, similarity between customers, etc. To experience the three areas making use of taoxnomic heterogeneity for adapting these, and to experience the techniques to use the taxonomy for analyzing customers preferences, this chapter is presented.

In detail, this work aims at providing a comprehensive elaboration of recent progress in the areas of catalog segmentation, dynamic taxonomies including faceted search, and personalized directories. For each of the mentioned research areas, recent techniques are discussed, as well as the most recent real-world applications are investigated in detail. In addition, the area of recommender systems is also investigated in a comprehensive manner. Hereby, types and characteristics of existing methods are discussed, and the most recent and most promising concrete techniques are elaborated.

The remainder of this chapter at hand is organized as follows. In Sects. 5.1, 5.2, and 5.3, the research areas are explained aiming to evolve the taxonomy according to specific requirements. This includes the area of dynamic taxonomies using Sect. 5.1, the area of catalog segmentation using Sect. 5.2, and the area of personalized directories using Sect. 5.3. The area of recommender systems is discussed in Sect. 5.4, before the chapter concludes in Sect. 5.5.

5.1   Dynamic Taxonomies

A dynamic taxonomy is based on a (static) taxonomy, but prunes itself in response to a user query [149]. This paradigm has been proposed as a solution to combine navigation and querying, offering both expressivity and interactivity [60]. Formally, a ***Dynamic Taxonomy*** Θ_{dyn} is a subset of the initial taxonomy Θ, see Eq. 5.1:

$$\Theta_{dyn} = (\zeta, \{\Phi_{dyn} \subset \Phi\}, \{\Lambda_{dyn} \subset \Lambda\}), \tag{5.1}$$

where ζ is a keyword filtering for the concepts to remain in Φ_{dyn}, which is a subset of a partially ordered set of concepts of Φ, and Λ_{dyn} is a set connecting the remaining concepts of Θ_{dyn}. Faceted search represents a subordinated group of dynamic taxonomies [169]. The origin of this approach lies in the keyword-based search scenario [39], where the input query is in the form of a set of keywords [19]. These methods combine text search with facets to help accessing documents metadata by refining the search results in an iterative way [12]. Each resulting facet includes information to support the current search query and is stored in a relational or hierarchical database [5]. The refinement is realized with partitioning the search space into orthogonal conceptual dimensions [70]. The grouping of the facets can be structured in three kinds of decision trees [16]: based on metadata, based on the number of results, or based on the choice of the searcher.

5.1.1 Dynamic Taxonomies Techniques

In dynamic taxonomies, the most relevant work exists in an adaptive faceted semantic browser for supporting the generation of dynamic facets [163]. The browser utilizes Semantic Web technologies, i.e., extension of the Web in the form of data formats and protocols, and an adaptation component that is based on an automatically acquired user model. For the latter, the browser exploits an external logging service for capturing events that occurred as a result of user interactions. As first, the relevance of facets for the individual users is determined based on his/her navigation history and facet selection. Dynamic facet generation is used, if the set of recently available facets is insufficient and thus does not meet the recent interests. Those dynamic facets can be added at run-time with respect to specific needs. The dynamic facets can be presented in three different ways. In the form of direct facets, where top-level facets are based on direct attributes, as nested facets, where the direct facet and similar facets are resulted, or in the form of indirect facets, where similar facets compared to the keyword are queried. This allows the user to refine the search query and for improving the required dynamic taxonomy. The adaptation process consists of four components.

1. *Ordering Facets.* All facets are structured descending regarding their relevance in the previous session.
2. *Selecting Active Facets.* The number of active facets is reduced to maximum three facets.
3. *Annotating Facets.* The facets of the active facet selection process are expressed with annotations describing the facet and the numbers of products that are connected.
4. *Emphasizing Facets.* The most relevant facets are highlighted in color.

The authors in [44] presented an approach for the visualization of dynamic facets in the form of topic pies. Their interface aims to help users for drilling down to relevant concepts, as well as supporting the user with an intuitive browsing. The topic pie itself consists of two levels, which are represented as rings: an outer-ring for displaying sub concepts and an inner-ring for visualizing the associated super concepts. The most benefit of this type of visualization lies in the equivalent representation of concepts, the enabling of navigating toward fine-grained topics, and the disambiguation of ambiguous search terms. The pie is adopted, each time a user searches for a corresponding topic. The required input for the algorithm consists of the search query itself, a predefined taxonomy, and a mapping function between both before-mentioned inputs. The algorithm works in three steps.

1. *Identifying the Query.* The first step aims in identifying the keyword search.
2. *Clustering the Concepts.* After weighting the keywords with the related concepts, the sub concepts are clustered into super concepts.
3. *Shrinking the Topic Pie.* Finally, the pies are limited with hiding irrelevant concepts.

5.1.2 Dynamic Taxonomies Application

As faceted search became a mainstream commercial technology of dynamic taxonomies during the last decade [162], this technique is indispensable for the e-commerce and retailer branches [167]. In e-commerce applications, the query over a faceted taxonomy is in the form of a filter [12], which is applied over the taxonomy. Therefore, the result is a subset of the taxonomy satisfying the desired filter [70]. This subset is described by a label and is also known as focus [24]. Facet ranking is necessary if the focus is too large. The ranking of the facets depends on the significance of the current query.

Besides the benefits of using faceted search, this paradigm subjects four major problems in e-catalogs. The first issue worth emphasizing is called the sparsity problem. It arises in the form of unexpected answers displayed to the user, because of incomplete document annotations [12]. The second problem concerns the lack of supporting heterogeneous domains. Most facets are generic and do not consider domain-specific terms [168]. The third problem is the lack of multiple classification, because the facets are derived from the metadata describing the products [73]. Furthermore, the benchmarks used for evaluating faceted search are predominantly based on subjective judgment, which leads to the fourth problem [169].

5.2 Personalized Directories

Since 2002, researchers effort to customize the expert generated taxonomy in the form of personalized directories, which are modified according to customers' requirements. Consequently, this paradigm has been proposed as a solution to overcome the inflexibility of the formal taxonomies. Formally, a **Personalized Directory** Θ_{per} is based on the initial taxonomy, but the set of concepts shown to the user are changed according to the customers' requirements, as well as the edges connecting the remaining concepts; see Eq. 5.2:

$$\Theta_{per} = (\kappa, \{\phi_{per}\}, \{\lambda_{per}\}), \tag{5.2}$$

where κ is an active customer, ϕ_{per} is a set of personalized concepts, and λ_{per} is a set of edges. Such techniques are often implemented using logic programming.

5.2.1 Personalized Directories in Logic Programming

In **Logic Programming**, e.g., with the logic programming language Prolog, the taxonomies as well as the system itself are not structured inside a database, but in the form of a knowledge base consisting of different predicates, namely facts, rules, as well as queries. Facts are predicates, which in contrast to rules, do not query other predicates inside rules for defining relationships between the facts [113]. Each fact (e.g., *concept(ID, Label, Super concept)*) includes a functor standing before the clinches, i.e., the name of the predicate (e.g., *concept*). In addition, a set of arguments captured between the clinches is used to describe the entity (e.g., *ID*, *Label*, and *Super concept*) [25]. The short form of representing predicates is to write the number of arguments behind the functor (e.g., *concept/3*). The arguments can be variables, which are starting with a capital letter, lists, atoms, numbers, or strings. A list (written as [...]) can include multiple atoms, strings, or other lists. A rule describes a collection of requirements that have to be fulfilled to unify a query, otherwise it fails. Each rule consists of a header and a body, which are separated with the if character (: −). Consequently, the header is also a predicate including arguments in the form of variables. A special form of the variable is called anonymous variable. It is written as underscore (_) and has no effect on other facts. The body itself can consist of other rules, facts, and regular expressions. Through this, the programming paradigm provides multiple benefits for knowledge management in e-commerce, especially for taxonomical engineering, including natural language processing, i.e., the transformation of unstructured data into structured data [65]. On the one hand, as logic approaches are geared to deal efficiently with larger sets of concepts, such techniques can deal with very large taxonomies (e.g., in biology), as well as with small business taxonomies. On the other hand, as logic approaches are exploiting the reasoning capabilities for automating tasks, this paradigm perfectly deals with tasks concerning the semantics of e-catalogs. And furthermore, as logic programming is cross-platform and database-independent, the frameworks implemented with this technique can be applied to all recently available standalone e-commerce applications, as well as on distributed information management systems.

The above-described benefits have been exploited across various applications in e-commerce. For example, the authors in [148] are proposing a on cognitive logic programming-based architecture to personalize recommendations. Hereby, the benefits are used to learn the strength and weakness of existing recommenders to provide more reliable and trustful recommendations. The authors in [124] are using logic programming to more efficiently control complex business rules for knowledge engineering. The comparison with other programming techniques has turned out that logical programming is more beneficial for business rules then the most comprehensive business rule manager Drools that is programmed using Java, which is an object-oriented programming language. And finally, logic programming has proven to improve Semantic Web applications. The authors in [41], for example, are presenting a logic-based tool named RDF Schema Explorer, which can parse, validate, query, and extend RDF Schemata over different collaboration-driven application domains.

5.2.2 *Personalized Directories Techniques*

The very first approach for generating personalized e-catalogs was introduced by [85]. This approach is limited to **B**usiness to **B**usiness (B2B) and aims in overcoming the problems of dynamically changing e-catalogs. Their approach follows the logic programming paradigm. For the modification of the taxonomies, five modification rules are presented: the spliting of a set of sub concepts into multiple sets, the shifting of a complete set of sub concepts to the level of the super concept, the displacement of a splited subset to a higher level, the shifting of a subset to a deeper level, and the assignment of new labels to the reduced super concepts. The explained modifications will be carried out if: a super concept has too less sub concepts, the depth of concepts is too deep, or a super concept has too many sub concepts. This conditions and the previously described rules have three goals in common: minimizing the depth of the taxonomy, balancing the initial tree, and minimizing the number of new labels. The semiautomatic approach performs in seven steps:

1. *Initializing Taxonomy.* Loading the recent structures of the taxonomies.
2. *Analyzing Taxonomy.* Computing the arithmetic/average mean and standard deviation of concepts per taxonomy.
3. *Selecting Concepts.* Selecting the concepts that have to be modified.
4. *Matching Taxonomies.* Matching the taxonomy against the conditions.
5. *Analyzing Candidates.* Computing the standard deviation of candidate taxonomies.
6. *Comparision.* Checking modification rules against each other for personalization.

7. *Iteration Process.* Repeats steps three to six until a satisfying modification is resulted.

A system for presenting e-catalogs based on OWL was introduced by [57]. Their main contribution is the development of an e-catalog management system that helps the experts as well as the customers to create, update, browse, present, classify, and customize their individual e-catalog. The query against the taxonomies is realized with utilizing SPARQL. It supports taxonomical, facet-based, as well as keyword-based search against the e-catalogs. New developed concepts are semiautomatically mapped against the standard taxonomy classification systems. Furthermore, a semantic recommendation procedure is included to recommend related products to the developed concepts. Hereby, the recommender component only recommends products, which the customer is most likely to buy to avoid poor recommendation. The proposed products can be rated through explicit feedback. The analysis of individual preferences is measured inside a matrix comparing each customer with his/her ratings for individual products. The matrix is updated after every purchasing process. Their recommendation procedure runs in five steps.

1. *Selecting Attributes.* The user starts with selecting individual labels for concepts.
2. *Applying WordNet.* Equivalences are measured with WordNet against the system.
3. *Determining Equivalences.* The concepts with the same attributes are determined.
4. *Measuring Similarity.* The similarity between the selected concepts and the current object is calculated by using Cosine similarity.
5. *Selecting Recommendations.* The user can choose the recommended concepts.

The most recent work on personalizing e-catalogs was introduced by [9]. Their main contribution is an expert systems that is capable to fully automatically adapt the taxonomy based on an included recommender system. The system is capable regarding the differentiation between long- and short-term preferences and hybrid preference analysis, which allows an accuracy similar to the expert manual modification. In contrast to all the other works on personalized directories, and the related areas of dynamic taxonomies and catalog segmentation, the provided modification rules do not reduce the taxonomy semantically. This means that even if the number of concepts is reduced, still the same amount of information is included in the taxonomy. This is achieved through using so-called mediator concepts. The first type of mediator concept, named taxonomic dependencies, is used to combine sub concepts, and to allow the following modification rules: binding low preferred sub concepts to a single dependency, spliting middle preferred concepts to single sub concepts, and moving high preferred sub concepts to a higher level inside the taxonomy by moving the dependency to the level of a super concept. The second type of mediator concept, named taxonomic binding, is used to balance the taxonomy according to the changes performed on the taxonomic dependencies. The following modification rules are considered for bindings to overcome a semantically loss: binding all/none is performed if all dependencies belonging to a super concept is performed, and binding some means that the super concept is changed because of a dependency is moved to a higher level. For creating such mediator concepts the authors introduced a decision support system making extensively use of the advances in taxonomy matching, including the usage of a flexible matching strategy, the usage of various sources of background knowledge, as well as user involvement [10]. The taxonomy adaption process for finally adapting the taxonomy according to the customers' preferences based on logic programming performs in seven steps. Hereby, the first three steps are used to forecast the customers' requirements, and the four last steps are utilized to create the personal taxonomy:

1. *Past-term analysis.* This content-based is used for analyzing the customers' long- and short-term preferences.
2. *Future-term analysis.* This collaborative filtering technique makes use of the CRM taxonomy for analyzing similar customers.
3. *Hybrid aggregation.* Past-term preferences and future-term preferences are combined to finally predict upcoming preferences.
4. *Identification.* Based on the preferences, for the first type of mediator concepts (dependencies), the modification rules are performed.
5. *Binding all/none.* For super concepts not requiring a balance of the taxonomy, because of all dependencies are moved to a higher level or all dependencies remain as sub concepts of the super concept, either bound or split.
6. *Binding some.* For super concepts requiring a balance of the taxonomy, because not all dependencies remain as sub concepts of the super concept. Consequently, the super concept is renamed according to the sub concepts remaining.
7. *Output.* Based on the preferences and the performed modifation rules, the taxonomy can be output in different data formats.

5.2.3 Personalized Directories Application

Until now, the techniques of personalized directories have not affected the rank and file of real-world e-commerce applications. This is because the modification has to be performed manually, which is inefficient for high-traffic retailing markets, and because each modification is changing the semantics inside the taxonomy. For example, a not correct change of a super concept when moving sub concepts to a higher level would hamper the customers in finding the desired products, as the initial label of the super concept becomes inadequate.

One real-world approach for personalized directories was implemented by an Indian newspaper.[1] Hereby, the categories used during a navigation process are remembered, before moving the used categories to the top level of the directory.

---

[1] http://timesofindia.indiatimes.com/.

5.3 Catalog Segmentation

Techniques following the paradigm of catalog segmentation are assigning customers to predefined subcatalogs [173]. This paradigm has been proposed as a solution to effectively reduce the e-catalog according to the customers' preferences. Formally, a *Segmented Catalog* Θ_{seg} is a subset of the initial taxonomy Θ, see Eq. 5.3:

$$\Theta_{seg} = (\{K\}, \{\Phi_{seg} \subset \Phi\}, \{\lambda_{seg} \subset \Lambda\}), \qquad (5.3)$$

where K is a group of customers assigned to the segmented catalog Θ_{seg}, and Φ_{seg} is a subset of the initial set of concepts, and Λ_{seg} is a subset of the initial set of edges. Consequently, the design of segmented catalogs consists of the design of multiple catalogs, in which each catalog includes a different number of products most covering the customers needs assigned to a specific catalog [105].

5.3.1 Catalog Segmentation Techniques

The approach in [7] focuses on the design of e-catalogs, but from the side of the marketing expert, not from the customer's side. For the computation of preferences for a predefined set of catalogs they proposed two algorithms. The first algorithm removes the customers that are not interested in the predefined catalogs. Its basis is a list of all catalogs, each including all known customers. The analysis of preferences is made with the help of implicit knowledge or with explicit feedback. The second algorithm performs in a contrary way. Its basis is an empty set of customers filled when a user is interested in a predefined minimum number of products in the catalog. The preference value is limited to a Boolean operator, and thus can only be either true or false. Both explained algorithms above run in two steps.

1. *Initializing Taxonomies.* A catalog includes a subset of products and all customers.
2. *Rejecting Customers.* Irrelevant customers are removed from the catalog.

1. *Initializing Taxonomies.* A set of uncovered products is produced.
2. *Adding Customers.* Customers are added to the catalog if they are interested in a minimum number of products.

The system proposed in [102] aims in helping marketing managers to establish better marketing strategies. For mining the customer behavior, the system utilizes rules from database design. The proposed database can be grouped into three major components: the customer database, which stores data about the customer, the product database, which includes products and their metadata, and the stationery mall database, which includes a list of all physical departments. Those aim in supporting the marketing manager for distributing cross-selling strategies, and for the design of new e-catalogs. The mining system itself consists of six components: one component concerning the database establishment for collecting customer data and transaction data, the data mining component for analyzing consumer behaviors, the segmentation analysis for measuring segmentation and brand likings, the knowledge acquisition for building the knowledge bases, and the catalog marketing and sales promotion components for the distribution of marketing strategies. The data mining process inside the relational database requires four steps.

1. *Loading the Association Rules.* The association rules are loaded into the relational database.
2. *Analyzing the Preferences.* The customer behavior is mined with implicit feedback.
3. *Designing the Catalog.* Designing the catalog according to the preferences.
4. *Distributing the Catalog.* Distributing the e-catalog to the relevant customers.

5.3.2 Catalog Segmentation Application

In e-catalogs, the analysis of customer interactions and needs for a specific marketplace can be performed with data mining (i.e., process of discovering patterns from data sets), optimization methods, or combined approaches. In this area, the approaches differ in the techniques used to infer customers' preferences and in the ways the customer-oriented catalogs are created.

The most recent technique of catalog segmentation is utilizing customer knowledge provided inside a CRM system and assigns each covered customer to maximum one cluster of products [105].

5.4 Analyzing Customers' Requirements with Recommender Systems

Recommender systems attempt to provide solutions to analyze users particularly requirements, and for automatically selecting those items (e.g., concepts), which might be interesting for the analyzed customer [22]. It so helps guiding the individual users through the large range of products in e-commerce sites and changed the way that people are interacting with the Web [145]. The development of a recommender system is multidisciplinary as many research areas have to be combined [144]: artificial intelligence, information modeling, data mining, user interfaces, decision support systems (system that makes use of data(base) to suggest operations), marketing, and customer behavior. Furthermore, the choice of the right recommendation algorithms essentially depends on the type of products and the amount of available information [165]. Another important issue is the differentiation between more recent preferences, formally known as short-term preferences, and more distant preferences, known as long-term preferences. Both respond to the fact that user preferences usually change over time. The recommendation itself works in three major steps by which each task is usually handled in a separate component [104]:

1. *Pre-Processing Information.* The first step aims for structuring relevant information from text into keyword-based vectors.
2. *Learning User Profile.* The step for learning the user profile has the scope to collect information for deriving user preferences.
3. *Filtering Items.* The final task for filtering items intends to match the analyzed dislikes and likes against the items to be recommended.

Resulting is a matrix of preferences between concepts and customers, as given in Eq. (5.4), in which

$$cpref(\Theta) = \begin{pmatrix} \alpha_1, \alpha_2, \ldots \\ \kappa, \kappa, \ldots \\ \phi_1, \phi_2, \ldots \\ \rho, \rho, \ldots \end{pmatrix}, \tag{5.4}$$

where $cpref(\Theta)$ denotes a matrix of preferences ρ between the customer κ and a concept ϕ of taxonomy Θ. Recent recommender systems like Athena [80] or GroupLens [143] have proven to be useful in e-commerce applications [29], to suggest items and help determining which products to purchase [174], in social networks for finding friends, news, photos [165].

5.4.1 Types and Characteristics of Recommendation Methods

Recommendation methods can be distinguished in three major directions, namely collaborative filtering and content-based, as well as the combination of both, namely hybrid recommendation [103, 135].

Collaborative Filtering assumes that each customer belongs to a larger group of users with similar behavior [143]. During the last two decades, different approaches concerning collaborative filtering emerged. However, all this work can be classified into two classes of algorithms: memory-based or model-based [101]. The former stands for the initially aim of collaborative filtering [45] where two kinds of strategies can be distinguished: user-based algorithms on the one side and item-based algorithms on the other side. First mentioned approaches represent the user as a vector in a space of items [143], whereas the second mentioned approaches represent the items in the user space [151]. It hereby aims in finding the nearest neighbor inside the vector of items, respectively the user. Model-based algorithms aim in defining predictions. Two types of model-based algorithms represent the core research directions: aspect model approaches and personality diagnosis model approaches. While aspect model approaches are based on the mixture to model users ordering behavior, personality diagnosis model approaches are based on a training set [45]. Collaborative filtering overcomes the overspecialization problem, i.e., not cross-domain, but looses accuracy if the set of similar customers is too small, known as sparsity problem, or if no similar customers exist, known as cold-start problem [28, 135].

Content-Based recommendations suggest similar items to that ones the user preferred in the past [135]. It measures similarities between items based on textual information, e.g., product descriptions [165]. This recommendation method follows a two-step consisting process: the indexing of products, and the indexing of users. The latter is called profile learning and is based on implicit feedback, e.g., purchases, and explicit feedback, e.g., product reviews. Resulting is a set of keywords associated with the item, respectively the users profile [170]. Secondly, the comparison between the products and the users [170], to recommend items similar to those a given user has liked in the past [104]. One special kind of content-based recommendation is the method of *Knowledge-Based* recommendation. This technique adds implicit knowledge about the user, e.g., demographic information, to deduce appropriate recommendations [88]. Content-based recommendation has been proven for applications, which concentrate on the recommendation of documents relevant to a topic [99]. This approaches lack of the semantic understanding of users preference, and so the resulted recommendations only include items very similar to those the user already knows [135].

Hybrid-Based recommendation, as the name suggests, combines both previous-mentioned recommendation strategies to improve performance and to overcome their limitations [99].

5.4.2   Recommender Systems Techniques

The authors in [164] introduced a recommender system that is based on the knowledge captured inside a taxonomy. Their work aims at enhancing the effectiveness and accuracy with using an enriched representation of the semantics concerning contents in the space of the retrieval task. Furthermore, by combining retrieval results coming from short-term user preferences and long-term user preferences. Their framework consists of three components. A personalization system exploits metainformation about the user. The required information can either be provided explicitly or implicitly. The next component, the retrieval system, exploits the metainformation of the products to measure their similarity. A further component, the personalization system utilizes the user profile to represent preferences in the form of a vector. The values of preferences can be scaled, depending on the interests on products inside a taxonomic concept. The algorithm performs in five steps:

1. *Measuring Preferences.* Measuring the users' preferences.
2. *Comparing Conceptual Semantics.* Propagating the weights of semantic relations between single concepts.
3. *Expanding Relationships.* Expanding the context regarding the results of the semantic relationships inside the taxonomy.
4. *Extending Preferences.* Extending the preferences for users inside a vector.

5. *Contextualizing.* Combining the former steps for contextualizing the users' preferences.

A similar work was introduced by [38] and builds up on the work from [135]. It is a system for the direct recommendation of items and is also based on the analysis of conceptual preferences. The recommender combines the strategies of content-based and knowledge-based methods inside a semihybrid recommender system. The feedback is collected through implicit and explicit feedback. Explicit feedback in the form of identifying past ratings is furthermore adopted in the form of learning users rating parameters. The scope is hereby that different users use various rating strategies. The intensity of preferences for concepts consists of four elements: the level similarity between taxonomic concepts, the traversed sequence length, the users conceptual preferences, and the correlation intensity between the discovered nodes and the starting nodes. Their genetic algorithm is formalized into a linear matrix equation and includes three components:

1. *User Preference Analysis.* Combination of two different profiling strategies, namely an analysis for concept preferences, and an analysis for products inside these concepts.
2. *Recommendation Strategies.* This component has the aim to search for items related to the users preferences by utilizing SPARQL query language.
3. *Personalized Estimating Model.* The goal of the last component is to estimate the user ratings for each recommended item.

5.5 Conclusion

This chapter has presented the related areas of taxonomy matching, respectively the research areas of making use of taxonomic heterogeneity. In detail, a review was performed for the research areas introducing approaches aiming to adapt the taxonomy according to specific requirements, and a review was performed for techniques aiming to experience the customers' requirements by making use of the taxonomy.

The chapter started with explaining the research area of dynamic taxonomies. Hereby, the aim of this research area was discussed, as well as the most recent techniques and the most recent applications. Afterward, the area of catalog segmentation was elaborated. Again, the field was discussed in detail, including the aim of the works published in this area, recent techniques, as well as best performing applications. Next, the paradigm of personalized directories was investigated. After defining this research area and recent advances, best performing techniques and current applications were explained. In addition, the basics of logic programming was explained, as many works of this area are using this programming paradigm. Finally, the area of recommender systems was discussed including an explanation of types and characteristics for recent techniques, and an investigation of most current techniques of this area.

Part IV
Conclusions

Chapter 6
Conclusions

Abstract The research presented in the previous chapters has addressed the recent state of the art for the research area of taxonomy matching. After discussing the background of taxonomy matching, the background of taxonomic heterogeneity has been discussed in detail. Here, the focus was given on explaining the aim of the research area, namely the matching between two heterogeneous taxonomies describing the same domain of interest. In addition, the problem when matching taxonomies has been explained in detail. Hereby, a novel methodology to accurately differentiate between the different types of heterogeneity has been presented. Afterward, the focus was on reviewing recent works in detail. After analyzing recent matching techniques, matching algorithms, and matching systems, matching evaluations, as well as matching datasets have been discussed. Here, the best performing attempts have been reviewed, as well as the most widely used evaluation campaign, named OAEI. Finally, related areas of research have been discussed in detail. Here, a review was presented for techniques and most recent applications in dynamic taxonomies, catalog segmentation, personalized directories, and recommender systems. Using the chapter at hand, the conclusions for the book are given.

The research presented in this book has addressed the recent advances in taxonomy matching including its related areas. In this chapter, the conclusions are explained and areas for works in the field of taxonomy matching are given.

For comprehensively reviewing the recent advances in the field of taxonomy matching, the book has used five chapters, which are as follows:

The book started in Chap. 1 with an informative summary of the research area of taxonomy matching. Hereby, the taxonomy principles have been explained, the aim of taxonomy matching, the problem of taxonomic heterogeneity, the categorization of matching attempts, as well as the metrics used for evaluation.

Chapter 2 has presented a methodology to more precisely evaluate the heterogeneity existing between taxonomies. In contrast to merely considering the four different types of heterogeneity, for each type its various levels of complexity have been examined. This was achieved by considering different degrees for terminological

© Springer International Publishing AG 2017

H. Angermann and N. Ramzan, *Taxonomy Matching Using Background Knowledge*, https://doi.org/10.1007/978-3-319-72209-2_6

heterogeneity, conceptual heterogeneity, syntactical heterogeneity, as well as for semiotic heterogeneity. Through using this methodology, it has been revealed for each type of heterogeneity, which level of complexity should strongly be considered for future research. Concretely, it has helped to critically analyze open issues for taxonomy matching attempts that aim to overcome taxonomic heterogeneity.

In Chap. 3, available matching techniques as well as recent matching algorithms and matching systems have been reviewed. In contrast to other literature reviews in the field, the review performed in this book has put particular importance to the core components of the matching systems, namely by reducing the approaches to its implemented matching strategies. Through this, the algorithms and systems are reviewed based on an effective combination of various techniques and the utilized sources of background knowledge. The different matching techniques, as well as the discussed algorithms and systems, have been continuously presented in a way using concrete examples and including comprehensive discussions.

In Chap. 4, the Ontology Alignment Evaluation Initiative campaigns between 2011 and 2015 were investigated. This comprehensive comparison of the progress has achieved three results concerning future work. Firstly, it has revealed, which techniques should be used and combined inside flexible matching strategies to assess similarity between concepts. In addition, it has pointed out, which alternative resources of background knowledge should be utilized to further help assessing semantic similarity and for logically highly increasing the matching quality result. Finally, through comparing the progress for the different tracks and test cases, it has evident the areas with highest potential for future research.

Using Chap. 5, related research areas have been discussed, which are making use of taxonomic heterogeneity. In detail, four related areas have been analyzed in detail. Firstly, the area of dynamic taxonomies that makes use of keywords to intuitively reduce the taxonomy. Secondly, recent research on personalized directories has been discussed that is aiming to adopt the taxonomy according to preferences. Thirdly, works on catalog segmentation has been reviewed aiming to create sub taxonomies based on preferences. Finally, recent progress in recommender systems has been analyzed providing different techniques to filter users preferences based on taxonomy and taxonomic heterogeneity.

The presented work on taxonomy matching has shown improvements and modifications. However, there are number of uncovered interesting topics that need to be addressed in future work: efficient matching and matching with background knowledge, user involvement and personalization, channel- and device-specific requirements, novel evaluation approaches, and some general directions for future work. For each topic of future work, the concrete challenges are as follows:

Efficient matching and matching with background knowledge provide the following areas for future research: Recent work TaxoSemantics [9] provides three matching strategies, which are scalable according to the used matching technique(s) and resources of background knowledge integrated. TaxoSemantics for assessing and extending e-catalogs could be further improved by the following directions. Firstly, alternatively to WordNet, standard and upper-level taxonomies like eClass or GS1 could be used. Those provide a more specific view on concepts across e-commerce domains and provide a further description for the concepts. This would allow to enrich the matching process by integrating a further language-based analysis in addition to using the WordNet gloss. However, as those do not combine the complete knowledge in one taxonomy, but differentiate between different sectors, a further step to pre-process the background knowledge would be required, as many retailers do not only focus on a single sector. Secondly, as the single sequences of MWEs usually have a different semantically weight, this can be considered for the background-based strategy. For example, adjectives can be weighted lower than nouns, as the nouns mainly affect the similarity between concepts. And thirdly, as taxonomies suffer from semiotic heterogeneity, i.e., misinterpretation of concepts, implicit feedback could be used to further analyze the goodness of the semantics of the taxonomy. Feedback like search queries, click history, or browsing behavior could be used to assess if the customers assume the sub concepts in the identical super concepts as done by the expert, and how the customers consider the sibling concepts as similar.

User involvement and personalization should be strongly improved in future attempts: As user involvement promises high potential to increase the matching quality results, it can be expected that this optional matching step will be improved. Until now, not all systems are capable to provide techniques for supporting the user when validating or correcting the detected correspondences. In addition, it is impor-tant to evolve taxonomies according to customer' requirements like TaxoPublish [9]. TaxoPublish for generating personalized directories could be further improved through increasing the flexibility of the included recommender system. Until now, most of the systems is considered for customer-specific B2B retailing. However, a requirement analysis for supporting B2C retailing would allow to significantly increase the application area of the presented expert system. Same for group-specific taxonomies, where the preferences of customer groups should be analyzed, or for branch-specific taxonomies, where the sectors have to be analyzed. Another aspect for predicting customers requirements would be to analyze unstructured data pro-vided in social media. So-called social listening techniques are capable to analyze customers moods according to specific brands, branches, or products. This could be considered to personalize the taxonomies according to the customers contributions in social platforms.

Channel- and device-specific requirements have to be considered as future direc-tions: Until now, the modification rules are performed depending on the definition of the provider. However, to also use a recommender system inside the cloud-based environment would allow to figure out if the customer is a multi- or a single-channel customer. With the help of this insights, each channel can be considered for each customer separately, instead of only considering the adaptation process according to the limitations and features of the channel itself.

More evaluation approaches are required: Recent matching algorithms that aim to match syntactical heterogeneous taxonomies are focussing on a matching between taxonomies stored in relational databases and taxonomies stored in semantic lan-guages. However, the comparison of the OAEI methodologies has shown that there is recently no track provided to evaluate such approaches. In addition, evaluation approaches should also focus on other domains and highly extend interactive matching track and should not concern about syntactical heterogeneity.

Some of the other challenges are as follows:

- Freely available large-scale datasets to assess matching algorithms,
- Parallelization and scalability and of the matching,
- Automatic computation versus human validation,
- Reduction of user involvement when turning matches into mapping,
- Using machine learning techniques to improve the mapping process,
- Automatic identification of metrics according to scenario,
- Extemporaneous methods of Precision and Recall.

Glossary

Arithmetic/Average Mean Average value of a set of values.
Standard Deviation Variation of a set of data values.
F-Measure Harmonic mean of Precision and Recall.
Precision Measure of correctness.
Recall Measure of completeness.
Accuracy Measure of systematical errors.
Balanced Accuracy Comparison of sensitivity and specificity.
Sensitivity Proportion of correctly identified positives.
Specificity Proportion of correctly identified negatives.
Sesame Sesame RDF database.

© Springer International Publishing AG 2017
H. Angermann and N. Ramzan, *Taxonomy Matching Using Background Knowledge*, https://doi.org/10.1007/978-3-319-72209-2

References

1. S. Abiteboul, R. Hull, V. Vianu, *Foundations of Databases: The Logical Level* (Addison-Wesley, Boston, 1995)
2. R. Agrawal, A. Borgida, H. Jagadish, Efficient management of transitive relationships in large data and knowledge bases, in *Proceedings of the 1989 ACM SIGMOD International Conference on Management of Data (ACM), Paris, France* (1989), pp. 53–262
3. R. Agrawal, A. Somani, Y. Xu, Storage and querying of e-commerce data, in *Proceedings of the 27th International Conference on Very Large Data Bases (Morgan Kaufmann Publishers), Rome, Italy* (2001), pp. 149–158
4. J. Aguirre, B. Grau, K. Eckert, J. Euzenat, A. Ferrara, R. van Hague, others, Results of the ontology alignment evaluation initiative 2012, in *Proceedings of the 7th International Workshop on Ontology Matching, Boston* (2012), pp. 1–43 (CEUR Workshop Proceedings)
5. H. Al-Aqrabi, L. Liu, R. Hill, L. Cui, J. Li, Faceted search in business intelligence on the cloud, in *Proceedings of the 2013 IEEE International Conference on Green Computing and Communications and IEEE Internet of Things and IEEE Cyber, Physical and Social Computing (IEEE), Beijing, China* (2013), pp. 842–849
6. F. Almoqhim, D. Millard, N. Shadbolt. An approach to building high-quality tag hierarchies from crowdsourced taxonomic tag pairs, in *Proceedings of the 5th International Conference on Social Informatics (Springer), Cham, Switzerland* (2013), pp. 129–138
7. A. Amiri, Customer-oriented catalog segmentation: effective solution approaches. Decis. Support Syst. (Elsevier) **42**, 1860–1871 (2006)
8. H. Angermann, N. Ramzan, E-commerce with smartstore.NET. VisualStudio1 (PPEDV AG) **66**, 39–42 (2016)
9. H. Angermann, N. Ramzan, Taxopublish: towards a solution to automatically personalize taxonomies in e-catalogs. Expert Syst. Appl. (Elsevier) **66**, 76–94 (2016)
10. H. Angermann, N. Zeeshan, P. Ramzan, Taxosemantics: assessing similarity between multi-word expressions for extending e-catalogs. Decis. Support Syst. (Elsevier) **98**, 10–25 (2017)
11. M. Arenas, A. Bertails, E. Prud, J. Sequeda, A Direct Mapping of Relational Data to RDF, Technical report, World Wide Web Consortium, 2012
12. M. Arenas, B. Cuenca-Grau, E. Kharlamov, S. Marciuska, D. Zheleznyakov, E. Jimenez-Ruiz, SemFacet: semantic faceted search over YAGO, in *Proceedings of the Companion Publication of the 23rd International Conference on World Wide Web Companion (IW3C2), Seoul, Korea* (2014), pp. 123–126
13. NAICS Association, North American Industry Classification System, Rockaway (NJ), vol. 16, pp. 1–771 (2007)

© Springer International Publishing AG 2017

H. Angermann and N. Ramzan, *Taxonomy Matching Using Background Knowledge*, https://doi.org/10.1007/978-3-319-72209-2

14. M. Atencia, A. Borgida, J. Euzenat, C. Ghidini, L. Serafini, A formal semantics for weighted ontology mappings, in *Proceedings of the 11th International Semantic Web Conference (Springer), Boston, Massachusetts, USA* (2012), pp. 17–33

15. D. Bailey, An efficient euclidean distance transform, in *Proceedings of the 11th International Semantic Web Conference (Springer), Berlin, Germany* (2004), pp. 394–408

16. S. Basu-Roy, H. Wang, G. Das, U. Nambiar, M. Mohania, Minimum-effort driven dynamic faceted search in structured databases, in *Proceedings of the 17th ACM Conference on Information and Knowledge Management (ACM), Napa Valley, California, USA* (2008), pp. 13–22

17. Z. Bellahsene, A. Bonifati, F. Duchateau, Y. Velegrakis, On evaluating schema matching and mapping, *Schema Matching and Mapping* (Springer, Berlin, 2011), pp. 253–291

18. D. Benz, A. Hotho, G. Stumme, Semantics made by you and me: self-emerging ontologies can capture the diversity of shared knowledge, in *Proceedings of the 2nd Web Science Conference (The Web Science Trust), Raleigh, North Carolina, USA* (2010), pp. 1–8

19. B. Bislimovska, A. Bozzon, M. Brambilla, P. Fraternali, Textual and content-based search in repositories of web application models. Trans. Web (ACM) **8**, 11:1–11:47 (2014)

20. C. Bizer, D2R MAP – a database to RDF mapping language, in *Proceedings of the 12th International World Wide Web Conference (ACM), Budapest, Hungary* (2003), pp. 1–2

21. C. Bizer, T. Heath, T. Berners-Lee, Linked Data - The Story So Far, Wright State University, Fairborn, Ohio, USA, 2009

22. Y. Blanco-Fernandez, J. Pazos-Arias, A. Gil-Solla, M. Ramos-Cabrer, M. Lopez-Nores, J. Garcia-Duque, others, A flexible semantic inference methodology to reason about user preferences in knowledge-based recommender systems. Knowl. Based Syst. (Elsevier) **21**, 305–320 (2008)

23. K. Bollacker, C. Evans, P. Paritosh, T. Sturge, J. Taylor, Freebase: a collaboratively created graph database for structuring human knowledge, in *Proceedings of the 2008 ACM SIGMOD International Conference On Management Of Data, Vancouver, British Columbia, Canada* (2008), pp. 1247–1250

24. D. Bonino, F. Corno, L. Farinetti, FaSet: a set theory model for faceted search, in *Proceedings of the 2009 IEEE/WIC/ACM International Joint Conference on Web Intelligence and Intelligent Agent Technology (IEEE), Milan, Italy* (2009), pp. 474–481

25. M. Bramer, *Logic Programming with Prolog* (Springer, Heidelberg, 2014)

26. D. Brickley, R. Guha, Resource description framework (RDF) schema specification, Technical report, World Wide Web Consortium, 2000

27. J. Brown, R. Dant, The role of e-commerce in multi-channel marketing strategy, *Handbook of Strategic E-Business Management* (Springer, Heidelberg, 2014), pp. 467–487

28. V. Busquet, L. Ceccaroni, Design, development and deployment of an intelligent, personalized recommendation system, Master dissertation, Universitat Politecnica de Catalunya, 2009

29. F. Cacheda, V. Carneiro, D. Fernandez, V. Formoso, Comparison of collaborative filtering algorithms: limitations of current techniques and proposals for scalable, high-performance recommender systems. Trans. Web (ACM) **5**, 2:1–2:33 (2011)

30. D. Calvanese, B. Cogrel, S. Komla-Ebri, R. Kontchakov, D. Lanti, M. Rezk, others, Ontop: answering sparql queries over relational databases. Semant. Web **8**(3(Preprint)), 1–17 (2016)

31. I. Cantador, I. Konstas, J. Jose, Categorising social tags to improve folksonomy-based recommendations. Web Semant. (Elsevier) **9**, 1–15 (2011)

32. M. Cheatham. Mapsss: results for oaei 2012. In *Proceedings of the 7th International Workshop on Ontology Matching, Boston* (2012), pp. 1–6 (CEUR Workshop Proceedings)

33. M. Cheatham, StringsAuto and MapSSS results for OAEI 2012, in *Proceedings of the 7th International Workshop on Ontology Matching (CEUR Workshop Proceedings), Boston, Massachusetts, USA* (2013)

34. M. Cheatham, Z. Dragisic, J. Euzenat, D. Faria, A. Ferrara, G. Flouris, others, Results of the ontology alignment evaluation initiative 2015, in *Proceedings of the 10th International Workshop on Ontology Matching (CEUR Workshop Proceedings), Bethlehem, Pennsylvania, USA* (2015), pp. 1–56

35. M. Cheatham, Z. Dragisic, J. Euzenat, D. Faria, A. Ferrara, G. Flouris, others, Results of the ontology alignment evaluation initiative 2015, in: *Proceedings of the 10th International Workshop on Ontology Matching, Bethlehem, PA, US* (2016), pp. 60–115

36. M. Cheatham, P. Hitzler, String similarity metrics for ontology alignment, *Semantic Web-ISWC 2013* (Springer, Heidelberg, 2013), pp. 294–309

37. J. Chen, D. Warren, Cost-sensitive learning for large-scale hierarchical classification, in *Proceedings of the 22nd ACM International Conference on Information and Knowledge Management* (2013)

38. S. Cheng, C. Chou, G. Horng, The adaptive ontology-based personalized recommender system. Wirel. Pers. Commun. (Kluwer Academic Publishers) **72**, 1801–1826 (2013)

39. P. Chippimolchai, V. Wuwongse, C. Anutariya, Semantic query formulation and evaluation for XML databases, in *Proceedings of the 3rd International Conference on Web Information Systems Engineering (IEEE)* (2002), pp. 205–214

40. V. Christophides, D. Plexousakis, M. Scholl, S. Tourtounis, On labelling schemas for the semantic web, in *Proceedings of the 12th International Conference on World Wide Web (ACM), Budapest, Hungary* (2003), pp. 544–555

41. W. Conen, R. Klapsing, Exchanging semantics with RDF, *Information Age Economy* (Springer, Heidelberg, 2001), pp. 473–486

42. I. Cruz, F. Antonelli, C. Stroe, Agreementmaker: efficient matching for large real-world schemas and ontologies. Proc. VLDB Endow. **2**, 1586–1589 (2009)

43. I. Cruz, C. Stroe, F. Caimi, A. Fabiani, C. Pesquita, F. Couto, others, Using the agreementmaker to align ontologies for the OAEI campaign 2007, in *Proceedings of the 2nd International Workshop on Ontology Matching (CEUR Workshop Proceedings), Bonn, Germany* (2011), pp. 1–6

44. T. Deuschel, C. Greppmeier, B. Humm, W. Stille, Semantically faceted navigation with topic pies, in *Proceedings of the 10th International Conference on Semantic Systems (ACM), Leipzig, Germany* (2014), pp. 132–139

45. Y. Ding, X. Li, Time weight collaborative filtering, in *Proceedings of the 14th ACM International Conference on Information and Knowledge Management (ACM), Bremen, Germany* (2005), pp. 485–492

46. D. Dinh, J. Dos Reis, C. Pruski, M. Da Silveira, C. Reynaud, Identifying change patterns of concept attributes in ontology evolution, *The Semantic Web: Trends and challenges* (Springer, Heidelberg, 2014), pp. 768–783

47. W. Djeddi, M. Khadir, Ontology alignment using artificial neural network for large-scale ontologies. Int. J. Metadata Semant. Ontol. (Inderscience Publishers) **8**, 75–92 (2013)

48. W. Djeddi, M. Khadir, XMapGen and XMapSiG results for OAEI 2013, in *Proceedings of the 8th International Workshop on Ontology Matching (CEUR Workshop Proceedings), Sydney, Australia* (2013), pp. 1–8

49. W. Djeddi, M. Khadir, XMap++ results for OAEI 2014, in *Proceedings of the 9th International Workshop on Ontology Matching (CEUR Workshop Proceedings), Riva del Garda, Italy* (2014), pp. 1–7

50. Z. Dragisic, K. Eckert, J. Euzenat, D. Faria, A. Ferrara, R. Granada, others, Results of the ontology alignment evaluation initiative 2014, in *Proceedings of the 9th International Workshop on Ontology Matching, Riva del Garda* (2014), pp 1–44 (CEUR Workshop Proceedings Workshop Proceedings)

51. F. Duchateau, Z. Bellahsene, Designing a benchmark for the assessment of schema matching tools. Open J. Databases (OJDB) **1**(1), 3–25 (2014)

52. J. Euzenat, A. Ferrara, R. van Hauge, L. Hollink, C. Meilicke, A. Nikolov, others, Results of the ontology alignment evaluation initiative 2011, in *Proceedings of the 6th International Workshop on Ontology Matching (CEUR Workshop Proceedings), Bonn, Germany* (2011), pp. 1–29

53. J. Euzenat, M. Rosolu, C. Trojahn, Ontology matching benchmarks: Generation, stability, and discriminability. J. Web Semant. **21**(Special Issue on Evaluation of Semantic Technologies), 30–48 (2015)

54. J. Euzenat, P. Shvaiko, *Ontology Matching*, 1st edn. (Springer, Heidelberg, 2007)
55. J. Euzenat, P. Shvaiko, *Ontology Matching*, 2nd edn. (Springer, Heidelberg, 2013)
56. D. Faria, C. Martins, A. Nanavaty, A. Taheri, C. Pesquita, E. Santos, others, Agreementmakerlight results for OAEI 2014, in *Proceedings of the 9th International Workshop on Ontology Matching (CEUR Workshop Proceedings), Riva del Garda, Italy* (2014), pp. 1–8
57. H. Farsani, A. Nematbakhsh, Designing a catalog management system - an ontology approach. Malays. J. Comput. Sci. (Faculty of Compuer Science and Information Technology) **20**, 119–127 (2007)
58. C. Fellbaum, *WordNet: An Electronic Lexical Database* (MIT-Press, Cambridge, 1998)
59. A. Ferrara, A. Nikolov, F. Scharffe, Data linking for the semantic web. Int. J. Semant. Web Inf. Syst. **7**, 46–76 (2011)
60. S. Ferre, O. Ridoux, Logical information systems: from taxonomies to logics, in *Proceedings of the 18th International Conference on Database and Expert Systems Applications (IEEE), Regensburg, Germany* (2007), pp. 212–216
61. M. Finlayson, Java libraries for accessing the princeton WordNet: comparison and evaluation, in *Proceedings of the 7th Global Wordnet Conference, Tartu, Estonia* (2014), pp. 1–8
62. N. Friedman, D. Geiger, M. Goldschmidt, Bayesian network classifiers. Mach. Learn. (Kluwer Academic Publishers) **29**, 131–163 (1997)
63. E. Friedman-Hill, Jess, the rule engine for the java platform, 2008
64. M. Gastmeyer, Standard-Thesaurus Wirtschaft, Institut fuer Weltwirtschaft, Kiel, Germany, 1998
65. A. Gomez-Perez, M. Fernandez-Lopez, O. Corcho, *Ontological Engineering: with Examples From The Areas of Knowledge Management, E-commerce And The Semantic Web* (Springer, Heidelberg, 2006)
66. B. Grau, Z. Dragisic, K. Eckert, J. Euzenat, A. Ferrara, R. Granada, others, Results of the ontology alignment evaluation initiative 2013, in *Proceedings of the 8th International Workshop on Ontology Matching, Sydney* (2013), pp. 1–40 (CEUR Workshop Proceedings Workshop Proceedings)
67. N. Guarino, D. Oberle, S. Staab, What is an ontology?, *Handbook on Ontologies* (Springer, Heidelberg, 2009), pp. 1–17
68. B. Hamp, H. Feldweg, GermaNet – a lexical-semantic net for german, in *Proceedings of ACL workshop Automatic Information Extraction and Building of Lexical Semantic Resources for NLP Applications, Madrid, Spain* (1997), pp. 1–7
69. T. Hayamizu, M. Mangan, J. Corradi, J. Kadin, M. Ringwald, The adult mouse anatomical dictionary: a tool for annotating and integrating data. Genome Biol. (BioMed Central) **6**, 1–8 (2005)
70. P. Heim, J. Ziegler, Faceted visual exploration of semantic data, in *Proceedings of the 2nd IFIP WG 13.7 Conference on Human-computer Interaction and Visualization (Springer), Uppsala, Sweden* (2011), pp. 58–75
71. E. Hellinger, Neue Begruendung der Theorie quadratischer Formen von unendlich vielen Veraenderlichen. J. fuer die reine und angewandte Mathematik (De Gruyter) **136**, 210–217 (2009)
72. V. Henrich, E. Hinrichs, R. Barkey, Aligning word senses in germanet and the dwds dictionary of the german language, in *Proceedings of the 7th Global WordNet Conference, Estonia* (2014), pp. 63–70 (Association for Computational Linguistics)
73. M. Henry, S. Hampton, A. Endert, I. Roberts, D. Payne, MultiFacet: a faceted interface for browsing large multimedia collections, in *Proceedings of the 2013 IEEE International Symposium on Multimedia (IEEE), Anaheim, California, USA* (2013), pp. 347–350
74. S. Herting, Hertuda results for OAEI 2012, in *Proceedings of the 7th International Workshop on Ontology Matching (CEUR Workshop Proceedings), Boston, Massachusetts, USA* (2012), pp. 1–4
75. P. Heymann, H. Garcia-Molina, Collaborative creation of communal hierarchical taxonomies in social tagging systems, Technical Report 2006-10, Stanford InfoLab, April 2006

76. M. Horridge, S. Bechhofer, The OWL API: a Java API for OWL ontologies. Semant. Web (IOS Press) **2**, 11 (2011)

77. I. Horrocks, Owl: A description logic based ontology language, in *Proceedings of the 21st International Conference on Logic Programming, Berlin* (Springer, Berlin, 2005), pp. 1–4

78. I. Horrocks, P. Patel-Schneider, H. Boley, S. Tabet, B. Grosof, M. Dean, others, Swrl: a semantic web rule language combining owl and ruleml. W3C Memb. Submiss. **21**, 79 (2004)

79. J. Huber, T. Sztyler, J. Noessner, C. Meilicke, CODI results for OAEI 2012, in *Proceedings of the 7th International Workshop on Ontology Matching (CEUR Workshop Proceedings), Boston, Massachusetts, USA* (2012), pp. 1–6

80. W. IJntema, F. Goossen, F. Frasincar, F. Hogenboom, Ontology-based news recommendation, in *Proceedings of the 2010 EDBT/ICDT Workshops (ACM), Lausanne, Switzerland* (2010), pp. 16:1–16:6

81. E. Ioannou, N. Rassadko, Y. Velegrakis, On generating benchmark data for entity matching. J. Data Semant. **2**(1), 37–56 (2013)

82. E. Jimenez-Ruiz, B. Grau, Logmap: Logic-based and scalable ontology matching, in *Proceedings of the 10th International Conference on The Semantic Web,Bonn* (Springer, Berlin, 2011), pp. 273–288

83. E. Jimenez-Ruiz, B. Grau, W. Xia, A. Solimando, X. Chen, V. Cross, others, LogMap family results for OAEI 2014, in *Proceedings of the 9th International Workshop on Ontology Matching (CEUR Workshop Proceedings), Riva del Garda, Italy* (2014), pp. 1–9

84. E. Jimenez-Ruiz, B. Grau, Y. Zhou, I. Horrocks, Large-scale interactive ontology matching: algorithms and implementation, in *Proceedings of the 20th European Conference on Artificial Intelligence (IOS Press), Montpellier, France* (2012), pp. 444–449

85. Y. Joh, J. Lee, Buyer's customized directory management over seller's e-catalogs: logic programming approach. Decis. Support Syst. (Elsevier) **34**, 197–212 (2003)

86. K. Jones, A statistical interpretation of term specificity and its application in retrieval. J. Doc. (Emerald Group Publishing) **28**, 11–20 (1972)

87. Y. Kang, L. Zhou, D. Zhang, An integrated method for hierarchy construction of domain-specific terms, in *Proceedings of the 2014 IEEE/ACIS 13th International Conference on Computer and Information Science (IEEE), Taiyuan, China* (2014), pp. 485–490

88. G. Karypis, Evaluation of item-based top-n recommendation algorithms, in *Proceedings of the 10th International Conference on Information and Knowledge Management (ACM), Atlanta, Georgia, USA* (2001), pp. 247–254

89. B. Kernighan, S. Lin, An efficient heuristic procedure for partitioning graphs. Bell Syst. Tech. J. (Blackwell Publishing) **49**, 291 (1970)

90. C. Kiu, E. Tsui, Taxofolk: a hybrid taxonomy-folksonomy structure for knowledge classification and navigation. Expert Syst. Appl. (Elsevier) **38**, 6049–6058 (2011)

91. S. Klai, M. Khadir, Approach for a rule based ontologies integration, in *Proceedings of the 5th International Conference on Computer Science and Information Technology, Yerevan* (2013), pp. 65–70 (Institute of Electrical and Electronics Engineers)

92. I. Kuo, T. Wu, ODGOMS results for OAEI 2013, in *Proceedings of the 8th International Workshop on Ontology Matching (CEUR Workshop Proceedings), Sydney, Australia* (2013), pp. 1–8

93. P. Lambrix, R. Kaliyaperumal, A session-based approach for aligning large ontologies, *The Semantic Web: Semantics and Big Data* (Springer, Berlin, 2013)

94. L. Lamport, *Document Preparation System* (Addison-Wesley, Boston, 1986)

95. C. Lee, Some properties of nonbinary error-correcting codes. Trans. Inf. Theory (IEEE) **4**, 77–82 (1958)

96. W. Levenshtein, Binary codes capable of correcting deletions, insertions and reversals. Sov. Phys. Dokl. (Springer) **10**, 707–710 (1966)

97. M. Ley, Dblp.uni-trier.de: Computer science bibliography

98. J. Li, J. Tang, Y. Li, Q. Luo, RiMOM: a dynamic multistrategy ontology alignment framework. Trans. Knowl. Data Eng. (IEEE) **21**, 1218–1232 (2009)

99. L. Li, W. Hong, T. Li, Taxonomy-oriented recommendation towards recommendation with stage, in *Proceedings of the 14th Asia-Pacific International Conference on Web Technologies and Applications (Springer), Kunming, China* (2012), pp. 219–230

100. S. Li, A. Jain, *Encyclopedia of Biometrics* (Springer, Heidelberg, 2009)

101. H. Liang, Y. Xu, Y. Li, R. Nayak, X. Tao, Connecting users and items with weighted tags for personalized item recommendations, in *Proceedings of the 21st ACM Conference on Hypertext and Hypermedia (ACM), Toronto, Ontario, CA* (2010), pp. 51–60

102. C. Lin, C. Hong, Using customer knowledge in designing electronic catalog. Expert Syst. Appl. (Elsevier) **34**, 119–127 (2008)

103. L. Liu, F. Lecue, N. Mehandjiev, Semantic content-based recommendation of software services using context. Trans. Web (ACM) **7**, 17:1–17:20 (2013)

104. P. Lops, M. de Gemmis, G. Semeraro, Recommender systems handbook, *Content-based Recommender Systems: State of the Art and Trends* (Springer, Heidelberg, 2011)

105. I. Mahdavi, M. Movahednejad, F. Adbesh, Designing customer-oriented catalogs in e-CRM using an effective self-adaptive genetic algorithm. Expert Syst. Appl. (Elsevier) **38**, 631–639 (2011)

106. P. Mayr, M. Stempfhuber, A. Walter, Auf dem Weg zum wissenschaftlichen Fachportal – Modellbildung und Integration heterogener Informationssammlungen, in *Proceedings of the 27th DGI-Online Tagung (Deutsche Gesellschaft fuer Informatik), Frankfurt am Main, Germany* (2005), pp. 1–13

107. B. McBride, Jena: a semantic web toolkit. IEEE Internet Comput. (IEEE) **6**, 1–55 (2002)

108. M. McCandless, E. Hatcher, O. Gospodnetic, *Lucence in Action: Covers Apache Lucence 3.0*, 2nd edn. (Manning-Publications, Greenwich, 2010)

109. D. McGuinness, F. Van Harmelen, OWL Web Ontology Language overview, Technical report, World Wide Web Consortium, 2004

110. C. Meilicke, R. Garcia-Castro, F. Freitas, W. Van Hage, E. Montiel-Ponsoda, R. De Azevedo, others, MultiFarm: a benchmark for multilingual ontology matching. Web Semantics (Elsevier) **15**, 62–68 (2012)

111. C. Meilicke, H. Stuckenschmidt, An efficient method for computing alignment diagnoses, in *Web Reasoning and Rule Systems*, ed. by A. Polleres, T. Swift (Springer, Heidelberg, 2009), pp. 182–196

112. S. Melnik, H. Garcia-Molina, E. Rahm, Similarity flooding: a versatile graph matching algorithm and its application to schema matching, in *Proceedings of the 18th International Conference on Data Engineering (IEEE), San Jose, California, USA* (2002), pp. 117–128

113. D. Merritt, *Building Expert Systems in Prolog* (Springer, Heidelberg, 1989)

114. G. Miller, WordNet: a lexical database for english. Commun. ACM (ACM) **38**, 39–41 (1995)

115. T. Mitchell, *Artificial neural networks, Machine Learning* (McGraw-Hill, New York, 1997)

116. M. Mohr, S. Russel, North american product classification system: concepts and process of identifying service products, in *Proceedings of the 17th Annual Meeting of the Voorburg Group on Service Statistics, Nantes, France* (2002), pp. 1–33

117. C. Mungall, C. Torniai, G. Gkoutos, S. Lewis, M. Haendel, Uberon, an integrative multi-species anatomy ontology. Genome Biol. (BioMed Central) **13**, 1–20 (2012)

118. M. Musen, National Cancer Institute thesaurus, Encyclopedia of Systems Biology (Springer, Berlin, 2013), pp. 1492–1492

119. M. Nakatsuji, Y. Fujiwara, Linked taxonomies to capture users' subjective assessments of items to facilitate accurate collaborative filtering. Artif. Intell. (Elsevier) **207**, 52–68 (2014)

120. V. Nebot, R. Berlanga, Efficient retrieval of ontology fragments using an interval labeling scheme (elsevier). Inf. Sci. **179**, 4151–4173 (2009)

121. L. Nederstigt, D. Vandic, F. Frasincar, An automated approach to product taxonomy mapping in e-commerce, *Management Intelligence Systems* (Springer, Heidelberg, 2012), pp. 111–120

122. D. Ngo, Z. Bellahsene, YAM++(-) results for OAEI 2013, in *Proceedings of the 8th International Workshop on Ontology Matching (CEUR Workshop Proceedings), Sydney, Australia* (2013), pp. 1–8

123. N. Noy, M. Musen, The PROMPT suite: interactive tools for ontology merging and mapping. Int. J. Hum. Comput. Stud. (Academic Press) **59**, 983–1024 (2003)

124. L. Ostermayer, D. Seipel. Knowledge engineering for business rules in Prolog, in *Proceedings of the 26th Workshop on Logic Programming (CEUR Workshop Proceedings), Bonn, Germany* (2012), pp. 1–24

125. L. Otero-Cerdeira, F. Rodriguez-Martinez, A. Gomez-Rodriguez, Ontology matching: a literature review. Expert Syst. Appl. (Elsevier) **42**, 949–971 (2015)

126. L. Otero-Cerdeira, F.J. Rodríguez-Martínez, A. Gómez-Rodriiguez, Ontology matching: a literature review. Expert Syst. Appl. **42**, 949 (2015)

127. P. Papadimitriou, P. Tsaparas, A. Fuxman, L. Getoor, TACI: taxonomy-aware catalog integration. Trans. Knowl. Data Eng. (IEEE) **25**, 1643–1655 (2013)

128. S. Park, W. Kim, Ontology mapping between heterogeneous product taxonomies in an electronic commerce environment. Int. J. Electron. Commer. (M. E. Sharpe) **12**, 69–87 (2007)

129. H. Paulheim, WeSeE-Match results for OEAI 2012, in *Proceedings of the 7th International Workshop on Ontology Matching (CEUR Workshop Proceedings), Boston, Massachusetts, USA* (2012), pp. 1–7

130. M. Pazienza, M. Pennacchiotti, F. Zanzotto, Terminology extraction: an analysis of linguistic and statistical approaches, *Knowledge Mining* (Springer, Heidelberg, 2005), pp. 255–279

131. E. Peukert, J. Eberius, E. Rahm, Amc - a framework for modelling and comparing matching systems as matching processes, in *Proceedings of the 27th IEEE International Conference on Data Engineering* (2011)

132. E. Peukert, J. Eberius, E. Rahm, A self-configuring schema matching system, in *Proceedings of the 28th IEEE International Conference on Data Engineering (IEEE), Arlington, Virginia, USA* (2012), pp. 306–317

133. D. Phan, D. Vogel, A model of customer relationship management and business intelligence systems for catalogue and online retailers. Inf. Manag. (Elsevier) **47**, 69–77 (2010)

134. E. Pianta, L. Bentivogli, C. Girardi, MultiWordNet: developing an aligned multilingual database, in *Proceedings of the 1st International Conference on Global WordNet (Central Institute of Indian Languages)* (2002), pp. 293–302

135. P. Pin-Yu, W. Chi-Hsuan, H. Gwo-Jiun, C. Sheng-Tzong, The development of an ontology-based adaptive personalized recommender system, in *Proceedings of the International Conference On Electronics and Information Engineering (IEEE), Kyoto, Japan* (2010), pp. 76–80

136. C. Pinkel, C. Binnig, E. Kharlamov, P. Haase, IncMap: pay-as-you-go matching of relational schemata to OWL ontologies with IncMap, in *Proceedings of the 2013th International Conference on Ontology Matching (CEUR Workshop Proceedings), Sydney, Australia* (2013), pp. 225–228

137. C. Pinkel, C. Binnig, E. Jiménez-Ruiz, W. May, D. Ritze, M.G. Skjæveland, A. Solimando, E. Kharlamov, *in RODI: A Benchmark for Automatic Mapping Generation in Relational-to-Ontology Data Integration* (Springer International Publishing, Cham, 2015), pp. 21–37

138. D. Powers, Evaluation: from precision, recall and f-factor to roc, informedness, markedness and correlation. J. Mach. Learn. Technol. (Bioinfo Publications) **2**, 37–63 (2007)

139. E. Prud'hommeaux, A. Seaborne, SPARQL query language for RDF, Technical report, World Wide Web Consortium, 2008

140. J. Quinlan, Induction of decision trees. Mach. Learn. (Springer) **5**, 81–106 (1986)

141. R. Rada, H. Mili, E. Bicknell, M. Blettner, Development and application of a metric on semantic nets. Trans. Syst. Man Cybern. (IEEE) **19**, 17–30 (1989)

142. S. Raunich, E. Rahm, Towards a benchmark for ontology merging, in *On the move to meaningful internet systems: OTM Workshops, Springer, Heidelberg/Berlin, Germany* (2012), pp. 124–133

143. P. Resnick, N. Iacovou, M. Suchak, P. Bergstrom, J. Riedl, GroupLens: an open architecture for collaborative filtering of netnews, in *Proceedings of the 1994 ACM Conference on Computer Supported Cooperative Work (ACM), Chapel Hill, North Carolina, USA* (1994), pp. 175–186

144. F. Ricci, L. Rokach, B. Shapira, P. Kantor, *Recommender Systems Handbook* (Springer, Heidelberg, 2010)

145. J. Riedl, B. Smyth, Introduction to special issue on recommender systems. Trans. Web (ACM) **5**, 1:1–1:2 (2011)

146. D. Rinser, D. Lange, F. Naumann, Cross-lingual entity matching and infobox alignment in wikipedia. Inf. Syst. **38**(6), 887–907 (2013)

147. C.R. Rivero, I. Hernandez, D. Ruiz, R. Corchuelo, Benchmarking data exchange among semantic-web ontologies. IEEE Trans. Knowl. Data Eng. **25**, 1997 (2013)

148. J. Sabater-Mir, J. Cuadros, P. Garcia, Towards a framework that allows using a cognitive architecture to personalize recommendations in e-commerce, in *Proceedings of the 11th European Workshop on Multi-Agent Systems (CEUR Workshop Proceedings), Toulouse, France* (2013), pp. 3–17

149. G. Sacco, G. Nigrelli, A. Bosio, M. Chiarle, F. Luino, Dynamic taxonomies applied to a web-based relational database for geo-hydrological risk mitigation. Comput. Geosci. (Pergamon Press) **39**, 182–187 (2012)

150. I. Sag, T. Baldwin, F. Bond, A. Copestake, D. Flickinger, Multiword expressions: a pain in the neck for NLP, in *Proceedings of the 3rd International Conference on Computational Linguistics and Intelligent Text Processing (Springer), Mexico City, Mexico* (2002), pp. 1–15

151. B. Sarwar, G. Karypis, J. Konstan, J. Riedl, Item-based collaborative filtering recommendation algorithms, in *Proceedings of the 10th International Conference on World Wide Web (ACM), Hong Kong* (2001), pp. 285–295

152. L. Schriml, C. Arze, S. Nadendla, Y. Chang, M. Mazaitis, V. Felix, others, Disease ontology: a backbone for disease semantic integration. Nucleic Acids Res. (BioMed Central) **40**, 1071–1078 (2011)

153. K. Sengupta, P. Haase, M. Schmidt, P. Hitzler, Editing R2RML mappings made easy, in *Proceedings of the 12th International Semantic Web Conference (CEUR Workshop Proceedings), Sydney, Australia* (2013), pp. 101–104

154. J. Sequeda, M. Arenas, D. Miranker, On directly mapping relational databases to RDF and OWL, in *Proceedings of the 21st International Conference on World Wide Web (ACM), Lyon, France* (2012), pp. 649–658

155. C. Shao, L. Hu, J. Li, RiMOM-IM results for OAEI 2014, in *Proceedings of the 9th International Workshop on Ontology Matching (CEUR Workshop Proceedings), Riva del Garda, Italy* (2014), pp. 1–6

156. G. Shen, Y. Liu, F. Wang, J. Si, Z. Wang, Z. Huang, D. Kang, OMReasoner results for OAEI 2014, in *Proceedings of the 9th International Workshop on Ontology Matching (CEUR Workshop Proceedings), Riva del Garda, Italy* (2014), pp. 1–44

157. P. Shvaiko, J. Euzenat, Ontology matching: state of the art and future challenges. Trans. Knowl. Data Eng. (IEEE) **25**, 158–176 (2013)

158. P. Sneath, Numerical taxonomy, in *Bergey's Manual of Systematic Bacteriology*, ed. by D.J. Brenner, N.R. Krieg, J.T. Staley, G.M. Garrity (Springer, Heidelberg, 2005), pp. 39–42

159. L. Sommaruga, P. Rota, N. Catenazzi, Tagsonomy: easy access to web sites through a combination of taxonomy and folksonomy, *Advances in Intelligent Web Mastering* (Springer, Heidelberg, 2011), pp. 61–71

160. P. Tan, M. Steinbach, V. Kumar, *Introduction to Data Mining* (Addison-Wesley, Boston, 2005)

161. P. Thuy, N. Thuan, Y. Han, K. Park, Y. Lee, RDB2RDF: completed transformation from relational database into RDF ontology, in *Proceedings of the 8th International Conference on Ubiquitous Information Management and Communication, Siem Reap, Cambodia* (2014), pp. 1–7 (ACM)

162. D. Tunkelang, *Faceted Search* (Morgan & Claypool, San Rafael, 2009)

163. M. Tvarozek, M. Bielikova, Personalized faceted navigation for multimedia collections, in *Proceedings of the 2nd International Workshop on Semantic Media Adaptation and Personalization (IEEE), London, United Kingdom* (2007), pp. 104–109

164. D. Vallet, P. Castells, M. Fernandez, P. Mylonas, Y. Avrithis, Personalized content retrieval in context using ontological knowledge. Trans. Circuits Syst. Video Technol. (IEEE) **17**, 336–346 (2007)

165. E. Vargiu, A. Giuliani, G. Armano, Improving contextual advertising by adopting collaborative filtering. Trans. Web (ACM) **7**, 13:1–13:22 (2013)
166. P. Verhoef, P. Kannan, J. Inman, From multi-channel retailing to omni-channel retailing: introduction to the special issue on multi-channel retailing. J. Retail. (Elsevier) **91**, 174–181 (2015)
167. M. Voigt, A. Werstler, J. Polowinski, K. Meissner, Weighted faceted browsing for characteristics-based visualization selection through end users, in *Proceedings of the 4th ACM SIGCHI Symposium on Engineering Interactive Computing Systems (ACM), Copenhagen, Denmark* (2012), pp. 151–156
168. B. Wei, J. Liu, J. Ma, Q. Zheng, W. Zhang, B. Feng, DFT-extractor: a system to extract domain-specific faceted taxonomies from wikipedia, in *Proceedings of the 22nd International Conference on World Wide Web Companion (IW3C2), Rio de Janeiro, Brazil* (2013), pp. 277–280
169. B. Wei, J. Liu, Q. Zheng, W. Zhang, X. Fu, B. Feng, A survey of faceted search. J. Web Eng. (Rinton Press) **12**, 41–64 (2013)
170. D. Werner, N. Silva, C. Cruz, A. Bertaux, Using DL-reasoner for hierarchical multilabel classification applied to economical e-news, in *Proceedings of the Science and Information Conference (IEEE), London, United Kingdom* (2014), pp. 313–320
171. R. Wille, *Formal Concept Analysis as Mathematical Theory Of Concepts And Concept Hierarchies* (Springer, Heidelberg, 2005)
172. Z. Wu, M. Palmer, Verbs semantics and lexical selection, in *Proceedings of the 32nd Annual Meeting on Association for Computational Linguistics (ACM), Las Cruces, New Mexico* (1994), pp. 133–138
173. Z. Xu, D. Du, D. Xu, Improved approximation algorithms for the max-bisection and the disjoint 2-catalog segmentation problems. J. Comb. Optim. (Springer) **27**, 315–327 (2014)
174. L. Yu, M. Dong, R. Wang, Taxonomy for personalized recommendation service, in *Proceedings of the 2008 International Symposium on Electronic Commerce and Security (IEEE), Guangzhou City, China* (2008), pp. 657–660
175. B. Zapilko, B. Mathiak, Object property matching utilizing the overlap between imported ontologies, in *The Semantic Web: Trends and Challenges*, ed. by V. Presutti, C. D'Amato, F. Gandon, M. D'Aquin, S. Staab, A. Tordai (Springer, Heidelberg, 2014), pp. 737–751
176. H. Zhang, H. Yu, D. Xiong, Q. Liu, HHMM-based chinese lexical analyzer ICTCLAS, in *Proceedings of the 2nd SIGHAN workshop on Chinese language processing (Association for Computational Linguistics), Stroudsburg, Pennsylvania, USA* (2003), pp. 184–187
177. A. Zubiaga, V. Fresno, R. Martinez, A. Garcia-Plaza, Harnessing folksonomies to produce a social classification of resources. Trans. Knowl. Data Eng. (IEEE) **25**, 1801–1813 (2013)

Index

© Springer International Publishing AG 2017
H. Angermann and N. Ramzan, *Taxonomy Matching Using Background Knowledge*, https://doi.org/10.1007/978-3-319-72209-2